POSTHARVEST

POSTHARVEST

AN INTRODUCTION TO THE PHYSIOLOGY AND HANDLING OF FRUIT AND VEGETABLES

R.B.H. Wills, W.B. McGlasson, D.Graham,
T.H. Lee and E.G. Hall

An a**vi** Book
Published by Van Nostrand Reinhold
New York

Published in the U.S.A. by
Van Nostrand Reinhold
115 Fifth Avenue
New York, New York 10003

Distributed in Canada
Macmillan of Canada
Division of Canada Publishing Corporation
164 Commander Boulevard
Agincourt, Ontario M18 3C7. Canada

16 15 14 13 12 11 10 9 8 7 6 5 4 3 2 1

Library of Congress Cataloging-in-Publication
Data

Postharvest: an introduction to the physiology
and handling of fruit and vegetables/R.B.H. Wills
. . . [et al.]. – 3rd ed.
p. cm.
Includes index.
ISBN 0-442-23943-2
1. Fruit – Postharvest physiology. 2. Vegetables –
Postharvest physiology. 3. Fruit – Postharvest
technology. 4. Vegetables – Postharvest
technology. I. Wills. R.B.H.
SB360.P66 1989
634',045–dc20

89-32467
CIP

Printed in Hong Kong by
South China Printing Company (1988) Limited

Contents

Acknowledgements

It is a pleasure for the authors to acknowledge the constructive criticism of the original manuscript by Professor Leonard Morris, Department of Vegetable Crops, University of California, Davis, the advice of Dr S. Jacobs, The Royal Botanic Gardens, Sydney in compiling Appendix II, John G. Gellatley, Special Entomologist, Biological and Chemical Research Institute, Department of Agriculture NSW for expert help in preparing Table 26, Miss Cathy Smillie for preparation of the line drawings and Mr W.E. Rushton, Senior Technical Officer (Photographer), CSIRO Division of Food Processing, for expert preparation of photographic materials.

The authors are grateful to the following colleagues who assisted with the revision of specific sections of the book: Dr John H.B. Christian, Chief Research Scientist, CSIRO Division of Food Processing (irradiation of food); Mr Arthur R. Irving, Senior Experimental Scientist, CSIRO Division of Food Processing (measurement of temperature and humidity); Mr Kevin J. Scott, Senior Research Scientist, New South Wales Department of Agriculture (ethylene removal from storage atmospheres); Dr Alister K. Sharp, Principal Research Scientist, CSIRO Division of Food Processing (handling, packaging, storage and transport technology and instrumentation); Dr Neil L. Wade, Senior Research Scientist, New South Wales Department of Agriculture (principles of cooling); and Dr Brian L. Wild, Senior Research Horticulturist, New South Wales Department of Agriculture (postharvest pathology).

Preface to first and second editions

In 1977 the authors were asked to arrange a short international training-course on the principles and practice of the postharvest handling and storage of fresh fruit and vegetables. This book is the outcome of the preparations for that course and the perceived absence of a contemporary, introductory textbook in this area. The authors have emphasised principles and included some technical information and important reference material. Topics include structure and composition of plant parts that are consumed in the fresh state, their physiology and bio-chemistry, storage temperature, humidity and water loss, modified and controlled atmospheres, postharvest disorders and diseases, quality evaluation, packaging, and requirements in the design and operation of cool stores. The text is intended for use in tertiary courses at technical colleges and universities and as a useful guide for individuals who may be employed as technologists by farming com-panies, transport organisations and retailing organisations, for packing house managers, cool storage operators, nutritionists and advisers to governments. It should also be of value to the concerned consumer.

Preface to the third edition

Since the conception of this book some ten years ago there has been a growing appreciation of the importance of the correct handling and storage of fresh fruit and vegetables, and accredited coursework in this area of study is now included in the curriculums of tertiary teaching institutions throughout the world.

The need for such a book has been amply demonstrated by the considerable international demand from teaching institutions, and both governmental and commercial organizations. Readers have reported general satisfaction with its format and organization, and, as a consequence, in this latest edition the original format has been retained but, of course, new information has been incorporated in line with the latest available research. The reading lists at the end of chapters have also been expanded, and in some cases complementary sources cited as it is realised that not all readers will have access to well-stocked libraries.

1
Introduction

Commerce in fresh fruit and vegetables arises from the importance of these commodities in human diets. Man has kept these commodities in his diet to provide variety, taste, interest and aesthetic appeal, and to meet certain essential nutritional requirements. Ascorbic acid (vitamin C), for example, is the most important nutrient, because man is unable to synthesise it. Furthermore, vegetables and some fruits can be important supplementary sources of carbohydrates, minerals and protein. The possible beneficial effects of dietary fibre derived from fruit and vegetables are currently under scrutiny as part of a re-examination of human diets of Western societies with the object of minimizing some diseases considered to be related to an affluent life-style.

The postharvest physiology of fresh fruit and vegetables has in recent times become an important subdivision of both plant physiology and horticulture. The increased attention afforded postharvest horticulture has mainly been due to the realization that faulty handling practices after harvest can cause large losses of produce that required large inputs of labour, materials and capital to grow. Informed opinion now suggests that increased emphasis should be placed on conservation after harvest, rather than endeavouring to further boost crop production, as this would appear to offer a better return for the available resources of labour, energy and capital. To increase the effectiveness of conservation measures, more must be known about the nature and causes of losses, due to both wastage and reduced quality, between harvest and consumption, and more people need to be trained in postharvest horticulture.

Various authorities have estimated that 25 to 80 per cent of fresh fruit and vegetables produced are lost after harvest, although a recent FAO survey only serves to indicate how vague and incomplete are many of these estimates. In tropical regions, which include a large proportion of the developing countries, these losses can assume considerable economic and social importance. In developed countries, such as North America, Europe and Australasia, postharvest wastage of fresh produce is often just as serious. As the value of fresh produce may increase many times from the farm to the retailer, the economic consequences of wastage at any point along the chain are serious. When farms are located near towns and cities, faulty handling practices are often less important because the produce is usually consumed before serious wastage can occur. Even in the tropical regions, however, production of some staple commodities is seasonal, and there is a need to store produce to meet requirements during the off-season. In industrialized countries and in countries which encompass a wide range of climatic regions, fresh fruit and vegetables are frequently grown at locations remote from the major centres of population. Thousands of tonnes of produce are now transported daily over long distances both within countries and internationally. Fresh fruit and vegetables are important items of commerce, and there is a huge investment of resources in transport, storage and marketing facilities designed to maintain a continuous supply of these perishable commodities. Thus the main role of the postharvest technologist is to devise methods by which deterioration of produce is restricted as much as possible during the period between harvest and consumption. Biochemically and physiologically, the postharvest technologist is mainly concerned with slowing down the rate of respiration of produce.

Scientific endeavour in the field of postharvest horticulture is now recognized by several international organizations. The American Society for Horticultural Science has a Working Group in Postharvest Horticulture. A major part of the activities of Commission C2 (food science and technology) of the International Institute of Refrigeration is devoted to the science and technology of the storage and handling of fresh fruit and vegetables after harvest. Working groups on the postharvest physiology of vegetables and fruit operate within the International Society for Horticultural Science. Research activities in postharvest horticulture are well established and during the last few years specialized courses to train technologists in this area have been introduced in several countries. In Australia tertiary programs in postharvest technology are offered by the University of New South Wales, the Queensland Agricultural College, and the Hawkesbury Agricultural College, N.S.W.

FURTHER READING

Coursey, D.G. Postharvest losses in perishable foods of the developing world. Lieberman, M. ed. Post-harvest physiology and crop preservation. New York: Plenum; 1983.

Coursey, D.G.; Proctor, F.J. Towards the quantification of post-harvest loss in horticultural produce. Acta Hortic. 49; 55–66; 1975.

Food and Agricultural Organization of the United Nations. An analysis of an FAO survey of post-harvest food losses in developing countries. Rome; 1977.

Kader, A.A. et al. Postharvest technology of horticultural crops. Berkeley, CA: University of California; 1985. Special Publication 3311.

National Academy of Sciences. Postharvest food losses in developing countries. Washington, DC; 1978.

2
Structure and composition of fruit and vegetables

DEFINITION OF FRUIT AND VEGETABLES

The botanical definition of a fruit—the product of determinate growth from an angiospermous flower or inflorescence—is too strict for the edible, fleshy fruits of commerce. The botanist's definition encompasses the fleshy fruits that arise from expansion of the ovary of the flower, and does not include fleshy fruits that arise from the growth of structures other than the ovary, such as the receptacle (apple, strawberry), bract and peduncle (pineapple). But it does include dry fruits, such as the nuts, grains and legumes, that are not commercially considered to be fruits. The *Shorter Oxford English Dictionary* defines fruit as 'the edible product of a plant or tree, consisting of the seed and its envelope, especially the latter when juicy and pulpy'. A consumer definition of fruit would be 'plant products with aromatic flavours, which are either naturally sweet or normally sweetened before eating': they are essentially dessert foods. These definitions are more suited to the common usage of the term fruit. The derivation of some common fruits from an ovary and surrounding tissues is shown in Figure 1. Most of the exaggerated developments of certain parts of the basic fruit structure arose naturally but have been accentuated by modern breeding programs to maximize the desirable features of each fruit and minimize the superfluous features. The production of seedless cultivars of certain fruits represents an extreme development in this latter respect.

The vegetables do not represent any specific botanical grouping, and exhibit a wide variety of plant structures. They can, however, be grouped into three main categories: seeds and pods; bulbs, roots and tubers; flowers, buds, stems and leaves. In many instances, the structure giving rise to the particular vegetable has been highly modified compared with that structure on the 'ideal' plant. The derivation of some vegetables is shown in Figure 2. The plant part that gives rise to the vegetable will be readily apparent when most vegetables are visually

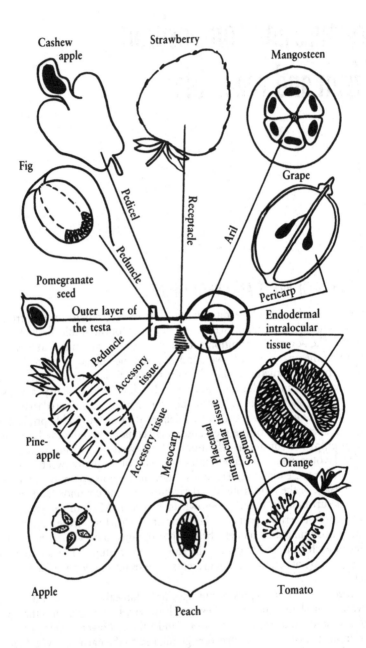

Figure 1 Derivation of some fruits from plant tissue. (Coombe, B.G. 'The Development of fleshy fruits.' *Annu. Rev. Plant Physiol.* 27, 1976, 507–28. With permission.)

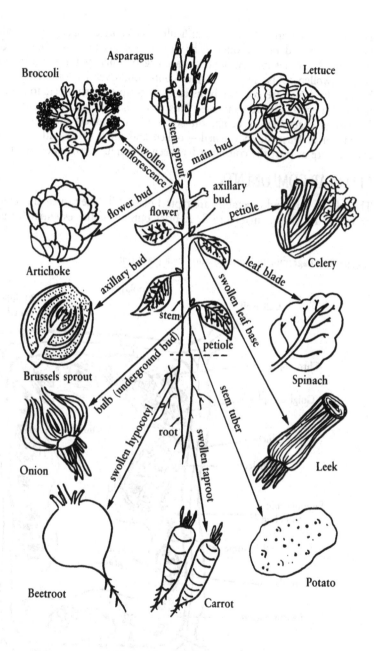

Figure 2 Derivation of some vegetables from plant tissue

examined. Some are a little more difficult to categorize, especially the tuberous organs developed underground. The potato, for instance, is a modified stem structure, but other underground storage organs, such as the sweet potato, are simply swollen roots. To add further confusion, several commodities that are botanically fruits are commonly listed as vegetables: cucumber, tomato, peas, beans and egg-plant are examples. This arises mainly from a consumer classification of vegetables as being soft edible plant products that are commonly salted—or at least not sweetened—cooked and often eaten with meat or fish dishes. Here at least the consumer has been triumphant over the scientist.

CELLULAR COMPONENTS

The cells of fruit and vegetables are typical plant cells, the principal components of which are shown in Figure 3. A brief outline of the essential features or

Figure 3 Diagrammatic representation of a plant cell

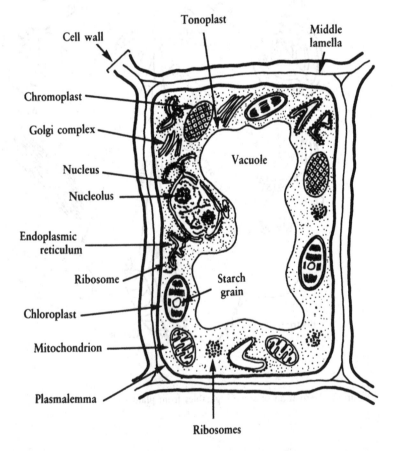

functions of these components will be given; more detailed explanations can be found in specialized texts such as those by Brouk and by Esau.

Plant cells are bounded by a more or less rigid cell wall composed of cellulose fibres, and other polymers such as pectic substances, hemicelluloses and lignins. A layer of pectic substances forms the middle lamella and acts to bind adjacent cells together. Adjacent cells often have small communication channels, called plasmadesmata linking their cytoplasmic masses. The cell wall is permeable to water and solutes. The main functions of the wall are:

1. to contain the cell contents by supporting the outer cell membrane, the plasmalemma, against the hydrostatic pressures of the cell contents which would otherwise burst the membrane; and
2. to give structural support to the cell and the plant tissues. Within the plasmalemma, the cell contents comprise the cytoplasm and usually one or more vacuoles. The latter are fluid reservoirs containing various solutes, such as sugars, amino and organic acids, and salts, and are surrounded by a semipermeable membrane, the tonoplast. Together with the semi-permeable plasmalemma, the tonoplast is responsible for maintaining the hydrostatic pressure of the cell, allowing the passage of water, but selectively restricting the movement of solutes or macromolecules, such as proteins and nucleic acids. The resulting turgidity of the cell is responsible for the crispness in fruit and vegetables.

The cytoplasm comprises a fluid matrix of proteins and other macromolecules and various solutes. Important processes which occur in this fluid part of the cytoplasm include the breakdown of storage reserves of carbohydrate by glycolysis (see Chapter 3) and protein synthesis. The cytoplasm also contains several important organelles, which are membrane-bound bodies with specialized functions as follows:

1. The nucleus, the largest organelle, is the control centre of the cell containing the genetic information in the form of DNA (deoxyribonucleic acid). It is bounded by a porous membrane that has distinct holes when viewed under the electron microscope. These permit the movement of mRNA (messenger ribonucleic acid), the transcription product of the genetic code of DNA, into the cytoplasm where mRNA is translated into proteins on the ribosomes of the protein synthesizing system (see below).
2. The mitochondria contain the respiratory enzymes of the tricarboxylic acid (TCA) cycle (see Chapter 3) and the respiratory electron transport system which synthesize adenosine triphosphate. Mitochondria utilize the products of glycolysis for energy production. Thus they form the energy powerhouse of the cell.
3. The chloroplasts, found in the green cells, are the photosynthetic apparatus of the cell. They contain the green pigment chlorophyll and the photochemical apparatus for converting solar (light) energy into chemical energy. As well, they have the enzymes neccessary for fixing atmospheric carbon dioxide to synthesize sugars and other carbon compounds.

4. The chromoplasts develop mainly from mature chloroplasts when the chlorophyll is degraded. They contain carotenoids which are the yellow-red pigments in many fruits.
5. Amyloplasts are the sites of starch grain development, although starch grains are also found in chloroplasts. Collectively chloroplasts, chromoplasts and amyloplasts are known as plastids.

Other membrane-bound systems within the cytoplasm are:

1. The Golgi complex, which is a series of plate-like vesicles that bud off smaller vesicles. These are probably of importance in cell wall synthesis and secretion of enzymes from the cell.
2. The endoplasmic reticulum is a network of tubules within the cytoplasm, which, some evidence suggests, may act as a transport system in the cytoplasm. What is more clear is that the endoplasmic reticulum often has attached to it ribosomes, which are the sites of protein synthesis. Other ribosomes are found free in the cytoplasm. The ribosomes contain ribonucleic acid and proteins.

CHEMICAL COMPOSITION AND NUTRITIONAL VALUE[1]

Water

Most produce contains more than 80 per cent water, with some tissues, such as cucumber, lettuce, marrow and melons, containing about 95 per cent water. The starchy tubers and seeds, for example yam, cassava and corn, contain less water, but even they usually comprise more than 50 per cent water. Quite large variations in water content can occur within a species, since the water content of individual cells varies considerably. The actual water content is dependent on the availability of water to the tissue at the time of harvest, so that the water content of produce will vary during the day if there are diurnal fluctuations in temperature. For most produce, it is desirable to harvest when the maximum possible water content is present as this results in a crisp texture. Hence the time of harvest can be an important consideration, particularly with leafy vegetables, which exhibit large and rapid variations in water content in response to changes in their environment.

Carbohydrates

Carbohydrates are generally the most abundant group of constituents. They can be present as low molecular-weight sugars or high molecular-weight polymers. They can account for 2–40 per cent of the tissue, with low levels being found in some cucurbits, for example, cucumber, and high levels in vegetables that accumulate starch, for example, cassava. Sugars are present mainly in ripe fruit, and starch occurs in both vegetables and unripe fruit. The main sugars present in fruit are sucrose, glucose and fructose with the predominant sugar varying in different

---

[1] All values are expressed on a fresh weight basis.

produce (Table 1). Glucose and fructose occur in all produce and are often present at a similar level.

Tropical and sub-tropical fruit tend to have the highest levels of glucose and fructose with persimmon, litchi, banana and pomegranate having combined levels of the two sugars of more than 10 per cent; grape is the only temperate fruit with more than 10 per cent. Sucrose is not present in all produce but occurs at 8–10 percent in the tropical fruits, rambutan, banana, carambola, mango and jackfruit and also in beetroot.

People can digest and utilize sugars and starch as energy sources, hence vegetables with a high starch content are important contributors to the daily energy requirement of people in many societies. Starch from plantain, cassava, yam, sweet potato and potato provides the bulk of energy in simple diets of subsistence groups in some developing countries. In these diets, over-dependence on the starchy vegetables is undesirable, as they cannot supply enough of the other essential nutrients. Consumers have a propensity for sugars as they produce the desired sweet taste in most fruits, but they also provide energy for the body.

Table 1 Sugar content of some ripe fruits[1]

| Fruit | Sugar (g 100 g fresh weight) | | |
	Glucose	Fructose	Sucrose
Apple	2	6	4
Banana	6	4	7
Cherry	5	7	0
Date	32	24	8
Grape	8	8	0
Orange (juice)	2	2	5
Peach	1	1	7
Pear	2	7	1
Pineapple	2	1	8
Tomato	2	1	0

[1] Adapted from Widdowson. E.M. McCance. R.A. The available carbohydrate of fruits. Determination of glucose, fructose, sucrose and starch. Biochem J. 29: 151 6: 1935.

A substantial proportion of carbohydrates is present as dietary fibre, which is not digested and passes through the intestinal system. Cellulose, pectic substances and hemicelluloses are the carbohydrate polymers that constitute fibre (Figure 4). Lignin, a complex polymer of aromatic compounds linked by propyl units, is also a major component of fibre. Dietary fibre is not digested by people as they are not capable of secreting the enzymes necessary to break down the polymers to the basic monomeric units that can be absorbed by the intestinal tract. Starch and cellulose have the same composition, as they are synthesized from D-glucose units, but the bonding between the monomers differs. Starch comprises α-1,4

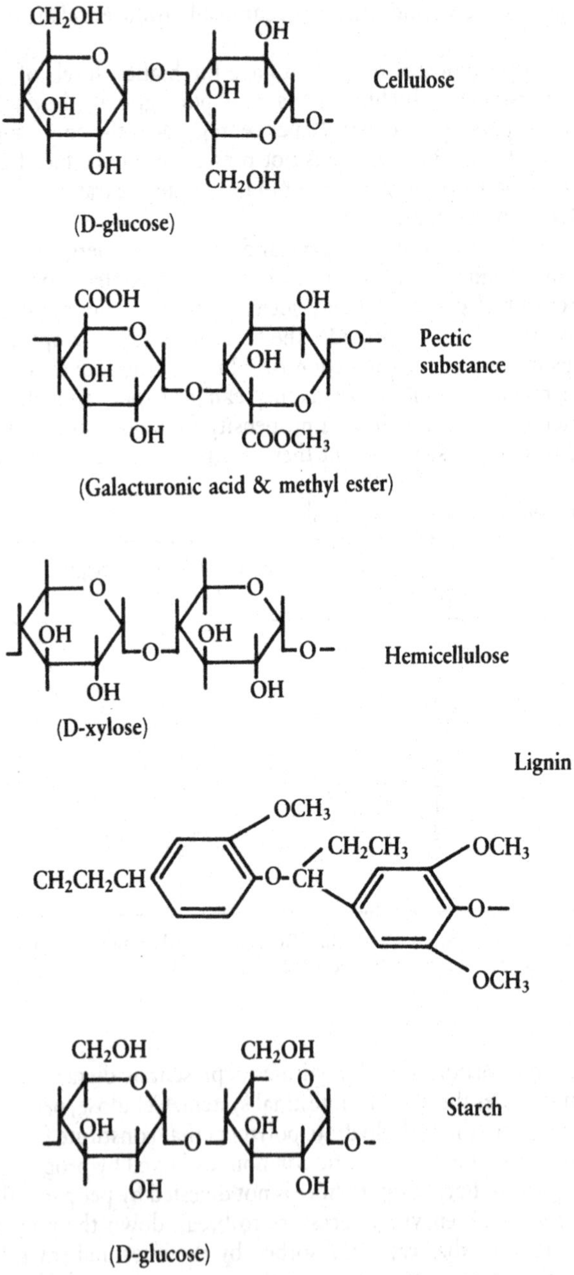

Figure 4 Structures of some fibre components and of starch

linkages, which are hydrolyzed by a range of amylase enzymes secreted by man; cellulose is formed with β-1,4 linkages, however, cellulose enzymes are not produced by man. Similarly man does not produce the necessary pectic enzymes and hemicelluloses to degrade respectively pectic substances to galacturonic acid units and hemicellulose to xylose and the other pentose constituents. Fibre was once considered to be an unnecessary component in the diet, although it was thought to relieve constipation. But fibre is now in vogue as the panacea to cure Western man of the diseases of civilization (Table 2). The evidence for the miracle properties of fibre are thus far only epidemiological and largely derived from comparisons between 'primitive' and 'civilized' societies. Further research is required to substantiate these claims.

Table 2 Diseases that have been claimed to be due to lack of fibre in diets

Appendicitis	Haemorrhoids
Cancer of colon	Hiatus hernia
Constipation	Ischaemic heart disease
Deep vein thrombosis	Obesity
Diabetes	Tumours of rectum
Diverticulosis	Varicose veins
Gallstones	

Protein
Fresh fruit and vegetables are not important contributors of protein to the diet. The protein content is generally about 1 per cent of fresh fruit and about 2 per cent in most vegetables, with the Brassica vegetables containing 3–5 per cent and the legumes containing about 5 per cent protein. The protein is mostly functional, for example, as enzymes, rather than a storage pool as in grains and nuts.

Lipids
Lipids comprise less than 1 per cent of most fruit and vegetables and are associated with protective cuticle layers on the surface of the produce and with cell membranes. The avocado and olive (used as a fresh vegetable) are exceptions, having respectively about 20 per cent and 15 per cent oil present as oil droplets in the cells.

Organic acids
Most fruit and vegetables contain organic acids at levels in excess of that required for the operation of the TCA cycle and other metabolic pathways. The excess is generally stored in the vacuole away from other cellular components. Lemon, lime, passionfruit and black currant often contain over 3 per cent of organic acids. The dominant acids in produce are usually citric and malic acid, and some examples are given in Table 3. Other organic acids that are dominant in certain commodities are tartaric acid in grapes, oxalic acid in spinach and isocitric acid in blackberries.

Table 3 Some fruits and vegetables in which citric and malic acids are the major acids present

Citric		Malic	
Berries	Beetroot	Apple	Broccoli
Citrus	Leafy vegetables	Banana	Carrot
Guava	Legumes	Cherry	Celery
Pear	Potato	Melon	Lettuce
Pineapple		Plum	Onion
Tomato			

Vitamins and minerals

Vitamin C (ascorbic acid) is only a minor constituent of fruit and vegetables but is of major importance in human nutrition for the prevention of the disease scurvy. Virtually all man's dietary vitamin C (approximately 90 per cent) is obtained from fruit and vegetables. The daily requirement of man for vitamin C is about 50 milligrams, and many commodities contain this amount of vitamin C in less than 100 grams of tissue.

Fruit and vegetables may also be important nutritional sources of vitamin A and folic acid, commonly supplying about 40 per cent of daily requirements. Vitamin A is required by the body to maintain the structure of the eye; a prolonged deficiency of vitamin A can eventually lead to blindness. The active vitamin A compound, retinol, is not present in produce, but some carotenoids such as β-carotene can be converted to retinol by man. Only about 10 per cent of the carotenoids known to be in fruit and vegetables are precursors of vitamin A; all other carotenoids such as lycopene, the main pigment in tomato, have no vitamin A activity. Folic acid is involved in RNA synthesis, and a deficiency will result in anaemia. Green leafy vegetables are rich sources of folic acid; the intensity of the green colour acts as a good guide to the folic acid content. Table 4 gives some important sources of vitamins C and A and of folic acid. Maintenance of these vitamins during handling and storage should be a major concern, particularly when the produce will be consumed by people on marginally sufficient diets.

The major mineral in fruit and vegetables is potassium and is present at more than 200 milligrams per 100 grams fresh weight in most produce. The highest levels are in green leafy vegetables with parsley containing about 1200 milligrams per 100 grams fresh weight but several vegetables contain 400–600 milligrams per 100 grams. Health authorities in many countries are urging increased consumption of potassium to counter the effects of sodium in the diet, and fruit and vegetables are the richest natural food source of potassium.

Table 4 Approximate levels of vitamin C, vitamin A and folic acid in some fruits and vegetables

Commodity	Vitamin C (mg 100 g)	Commodity	Vitamin A B-carotene equivalents (mg 100 g)	Commodity	Folic acid (μg 100 g)
Black currant, guava	200	Carrot	10.0	Spinach	80
Chili	150	Sweet potato		Broccoli	50
Broccoli, Brussels sprout	100	(red)	6.8	Brussels sprout,	
Papaya	80	Parsley	4.4	pulses	30
Kiwi fruit	70	Spinach	2.3	Cabbage, lettuce	20
Citrus, strawberry	40	Mango	2.4	Banana	10
Cabbage, lettuce	35	Red chili	1.8	Most fruits	< 5
Mango, carrot	30	Tomato	0.3		
Pineapple, banana, potato,		Apricot	0.1		
tomato, bean, cassava	20	Banana	0.1		
Apple, peach	10	Potato	0.0		
Beetroot, onion	5				

Many other vitamins and essential minerals are present in fruit and vegetables, but their contribution to total dietary requirements is generally of minor importance. Iron and calcium may be present at nutritionally significant levels, although often in a form that is unavailable for absorption by man, for example, most of the calcium in spinach is present as calcium oxalate which is not absorbed by man.

The nutritional value of various fruits and vegetables depends not only on the concentration of nutrients in the produce but also on the amount of such produce consumed in the diet. An attempt to equate these factors and show the relative concentration of ten major vitamins and minerals in some fruits and vegetables and their importance in the typical US diet is shown in Table 5. Tomatoes and oranges are relatively low in concentration of nutrients but make the major contribution of all produce to US diets because of the large per capita consumption.

Volatiles

All fruit and vegetables produce a range of small molecular-weight compounds (molecular weight less than 250) that possess some volatility at ambient temperatures. These compounds are not important quantitatively (normally less than 100 micrograms per gram fresh weight are present), but they are important in producing the characteristic flavour and aroma of fruit and to a lesser extent, of vegetables.

Table 5 Relative concentration of a group of ten vitamins and minerals in fruit and vegetables and the relative contribution of vitamins and minerals these commodities make to the US diet[1]

Nutrient Concentration		Contribution of nutrients to diet	
Crop	Rank	Crop	Rank
Broccoli	1	Tomato	1
Spinach	2	Orange	2
Brussels sprout	3	Potato	3
Lima bean	4	Lettuce	4
Pea	5	Sweet corn	5
Asparagus	6	Banana	6
Artichoke	7	Carrot	7
Cauliflower	8	Cabbage	8
Sweet potato	9	Onion	9
Carrot	10	Sweet potato	10
Sweet corn	12	Pea	15
Potato	14	Spinach	18
Cabbage	15	Broccoli	21
Tomato	16	Lima bean	23
Banana	18	Asparagus	25
Lettuce	26	Cauliflower	30
Onion	31	Brussels sprout	34
Orange	33	Artichoke	36

[1] Adapted from Rick, C.M. 'The tomato'. Sci. Am. 239(2): 66–76: 1978.

Table 6 Distinctive components of the aroma of some fruits and vegetables[1]

Product	Compound
Apple—ripe	Ethyl 2-methylbutyrate
—green	Hexanal, 2-hexenal
Banana—green	2-Hexenal
—ripe	Eugenol
—overripe	Isopentanol
Grapefruit	Nootakatone
Lemon	Citral
Orange	Valencene
Raspberry	1-(p-Hydroxyphenyl)-3-butanone
Cucumber	2,6-Nonadienal
Cabbage—raw	Allyl isothiocyanate
—cooked	Dimethyl disulphide
Mushroom	1-Octen-3-ol, lenthionine
Potato	2-Methoxy-3-ethyl pyrazine, 2.5-dimethyl pyrazine
Radish	4-Methylthio-*trans*-3-butenyl isothiocyanate

[1] Adapted from Salunkhe. D.K. Do. J.Y. Biogenesis of aroma constituents of fruits and vegetables CRC Crit. Rev. Food Sci. Nutr. 8: 161–90: 1977.

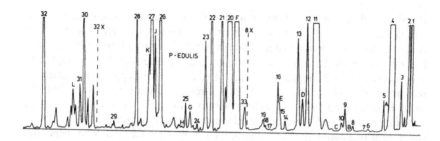

Figure 5   Gas chromatographic record of the volatile components of the juice of the purple passionfruit (*Passiflora edulis* Sims). Over forty compounds have been identified. Many are present in only small amounts, for example compounds located at A, B and E, but collectively they contribute to the characteristic aroma of passionfruit juice. The following is a list of some of the major components and a description of their aroma.[1]

Peak		
1	Ethyl acetate	Fruity, duco (paint) thinner
4	Ethyl butanoate	Pleasant tropical fruit, pineapple
11	Ethyl hexanoate	Pleasant tropical fruit, pineapple
12	*cis-* and *trans*-ocimene	Sweet floral, herbaceous
13	Heptyl acetate	Fruit, fatty-green
16	Hex-3-enyl acetate	Green fruity, passionfruit-skins
20	Hexyl butanoate and butyl hexanoate	Heavy fruity
21	Ethyl octanoate	Fruit-winey, reminiscent of apricot and banana
26	2-Heptyl hexanoate	Heavy fatty floral
30	Octyl hexanoate and hexyl octanoate	Sweet-fruity, orange-rose

Most fruits and vegetables each contain more than 100 different volatile compounds, mostly in minute amounts. The number of compounds known to be present in produce is continually increasing, as the sensitivity of the analytical techniques for their identification improves. The compounds are mainly esters, alcohols, acids and carbonyl compounds (aldehydes and ketones). Many of these compounds, such as ethanol, are common to all fruit and vegetables.

Definitive studies correlating consumer identification of the produce with the volatile profile emanating from the produce have shown that only a small number of compounds are responsible for consumer recognition of that commodity (Figure 5). In some fruits the characteristic aroma is due to the presence of one or two compounds. Table 6 gives the key compounds that have been claimed to be responsible for the characteristic aromas of some fruits and vegetables.

[1] Unpublished data—Dr F.B. Whitfield. CSIRO Division of Food Processing.

Practically all the compounds mentioned in Table 5 are minor components of the aroma fraction. The olfactory senses are thus extremely sensitive. The threshold concentration, or minimum concentration, at which the odour of ethyl 2-methyl-butyrate, the main characteristic odour of apple, can be detected organoleptically was found to be 0.001 microlitres per litre, that is, an apple of 100 grams is recognised if 0.01 micrograms of ethyl 2-methylbutyrate is present. For the characteristic odour to be desirable it must also be in the correct concentration. At different stages of maturation, different compounds become the dominant component of flavour so that a blindfolded subject would be able to detect the stage of development of a particular commodity by sniffing the aroma.

Details of the chemical composition of fruit and vegetables are to be found in many publications; much of the data from these sources are variable, because of inherent differences between cultivars and because of the effects of maturity, season and locality. The values presented here should, therefore, be regarded only as a guide to the chemical composition of fruit and vegetables.

FURTHER READING

Brouk, B. Plants consumed by man. London: Academic Press: 1975.

Davidson, S.; Passmore, R.; Brock, J.F.; Truswell, A.S. Human nutrition and dietetics. 6th ed. Edinburgh: Churchill Livingstone; 1975.

Duckworth, R.B. Fruit and vegetables. Oxford: Pergamon Press; 1966.

Esau, K. Anatomy of seed plants. 2d ed. New York: John Wiley and Sons; 1977.

Gebhardt, S.E.; Cutrifelli, R.; Matthews, R.H. Composition of foods; fruits and fruit juices, raw, processed, prepared. Washington, DC: US Department of Agriculture; 1982. Agriculture Handbook No. 8–9.

Haytowitz, D.B.: Matthews, R.H. Composition of foods: vegetables and vegetable products, raw, processed, prepared. Washington, DC: US Department of Agriculture; 1984. Agriculture Handbook No. 8–11.

Hulme, A.C., ed. The biochemistry of fruits and their products. Vols 1 and 2. London: Academic Press; 1970, 1971.

Johnson, A.E.: Nursten, H.E.; Williams, A.A. Vegetable volatiles: a survey of components identified. Parts I and II. Chem. Ind. 556–65, 1212–24; 1971.

McCarthy, M.A.; Matthews, R.H. Composition of foods: nut and seed products. Washington, DC: US Department of Agriculture; 1984. Agriculture Handbook No. 8–12.

Paul, A.A.; Southgate, D.A.T. McCance and Widdowson's the composition of foods. 4th revised ed. London: Her Majety's Stationery office; 1978.

Schreier, P. Chromatographic studies of biogenesis of plant volatiles. Heidelberg: Huthig; 1984.

Wills, R. B. H. Composition of Australian fresh fruit and vegetables. Food Technol. Aust. 39: 523–6; 1987.

3
Physiology and biochemistry of fruit and vegetables

A basic fact important in the postharvest handling of produce is that harvested fruit and vegetables are 'living' structures. One readily accepts that produce is a living, biological entity when it is attached to the growing parent plant in its agricultural environment. But even after harvest, the produce is still living as it continues to perform the metabolic reactions and maintain the physiological systems which were present when it was attached to the plant.

An important feature of plants and therefore of fruit and vegetables, is that they respire by taking up oxygen (O_2) and giving off carbon dioxide (CO_2) and heat. They also transpire, that is, lose water. While attached to the plant, the losses due to respiration and transpiration are replaced from the flow of sap, which contains water, photosynthates (principally sucrose and amino acids) and minerals. Respiration and transpiration continue after harvest, and since the produce is now removed from its normal source of water, photosynthates and minerals, the produce is dependent entirely on its own food reserves and moisture content. Therefore, losses of respirable substrates and moisture are not made up and deterioration has commenced. In other words, harvested fruit and vegetables are perishable.

This chapter will consider the postharvest behaviour of fruit and vegetables with particular reference to the physiological and biochemical changes that occur in ripening fruits. For this discussion, some understanding of the physiological development of fruit and vegetables is necessary.

PHYSIOLOGICAL DEVELOPMENT

The lives of fruit and vegetables can be conveniently divided into three major physiological stages following initiation or germination. These are growth, maturation, and senescence. However, clear distinction between the various stages is not easily made. Growth involves cell division and subsequent cell enlargement, which accounts for the final size of the produce. Maturation usually commences before growth ceases and includes different activities in different commodities. Growth and maturation are often collectively referred to as the development phase. Senescence is defined as the period when anabolic (synthetic) biochemical processes give way to catabolic (degradative) processes, leading to ageing and finally death of the tissue. Ripening, a term reserved for fruit, is generally considered to begin during the later stages of maturation and to be the first stage of senescence. The change from growth to senescence is relatively easy to delineate. Often the maturation phase is described as the time between these two stages, without any clear definition on a biochemical or physiological basis. It is difficult to assign specific biochemical or physiological parameters to delineate the various stages, because the parameters for different commodities are not identical in their nature or timing. Figure 6 shows the major stages in the life of the pineapple fruit, together with the changes in certain biochemical and physiological parameters. The relative changes in weight (as a measure of growth), chlorophyll, and flesh pH are common to many fruits, but for other parameters, such as the carotenoids and esters, the changes are specific for the pineapple. Special consideration will be given later to the different respiratory patterns by which fruit can be classified into climacteric and non-climacteric types. The pineapple is of the non-climacteric type. Analogous data to those for pineapple can be gleaned from the literature for many other commodities, although few individual researchers have accumulated complete sets of data for a particular commodity.

Development and maturation of fruit are completed only when it is attached to the plant, but ripening and senescence may proceed on or off the plant. Fruit are generally harvested either when mature or when ripe, although some fruits that are consumed as vegetables may be harvested even before maturation has commenced, for example, zucchini.

Similar terminology may be applied to the vegetables, or to any determinant organ, except that the ripening stage does not occur. As a consequence it is more difficult to delineate the change from maturation to senescence in vegetables. Vegetables are harvested over a wide range of physiological ages, that is, from a time well before the commencement of maturation through to the commencement of senescence (see Figure 36).

FRUIT RIPENING

The ripening fruit undergoes many physico-chemical changes after harvest that determine the quality of the fruit purchased by the consumer. Ripening is a dramatic event in the life of a fruit—it transforms a physiologically mature but inedible plant organ into a visually attractive olfactory and taste sensation. Ripening marks the completion of development (see Figure 6) of a fruit and the commencement of senescence, and it is normally an irreversible event. The following sections will discuss the general nature of fruit ripening, respiratory behaviour and the involvement of the gas ethylene (C_2H_4) with these processes.

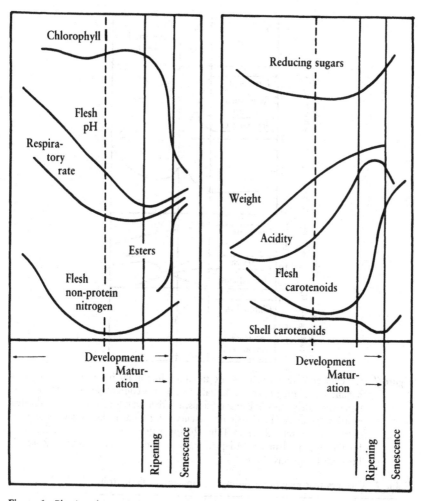

Figure 6   Physico-chemical changes in pineapple fruit during development. (Adapted from Gortner. W.A., Dull, G.G., Krauss, B.H. 'Fruit development, maturation, ripening, and senescence: a biochemical basis for horticultural terminology' HortScience 2; 141–4; 1967.)

Ripening is the result of a complex of changes, many of them probably occurring independently of one another. A list of the major changes which together make up fruit ripening are given in Table 7. The time course of some of these changes is shown in Figures 7 and 8 for banana and tomato, respectively, both of which are climacteric fruits. The principal difference between these fruits and the non-climacteric pineapple is the presence of the respiratory peak that is

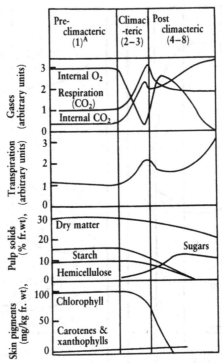

Figure 7   Generalized diagram to illustrate the more important biochemical changes in banana fruit during ripening. Actual values of gas concentrations and transpiration vary widely according to conditions, and their trends are, therefore, indicated in arbitrary units; the shapes of the curves for fruit in the overripe stage are extremely variable. Refer to Table 24 for a description of the colour stages for the ripening banana. (Adapted from Simmonds, N.W. Bananas, 2 nd ed. London: Longmans; 1966.)

Table 7   Changes that may occur during the ripening of fleshy fruit[1]

Seed maturation
Colour changes
Abscission (detachment from parent plant)
Changes in respiration rate
Changes in rate of ethylene production
Changes in tissue permeability
Softening: changes in composition of pectic substances
Changes in carbohydrate composition
Organic acid changes
Protein changes
Production of flavour volatiles
Development of wax on skin

---

[1] Adapted from Pratt. H.K. The role of ethylene in fruit ripening. Facteurs et régulation de la maturation des fruits. Anatole. France. Centre National de la Recherche Scientifique: 1975: 153–160.

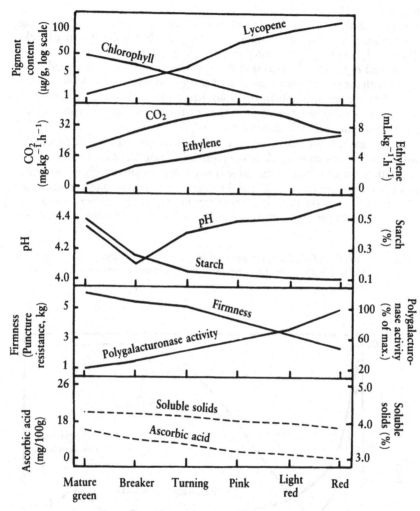

Figure 8　Some physico-chemical changes that occur during the ripening of tomato fruit. (Adapted from Rick, C.M. The tomato. Sci. Am. 239 (2): 66–76; 1978).

characteristic of climacteric fruits. A sharp increase in respiration is shown by the increase in production of carbon dioxide or decrease in internal oxygen concentration. Two of the changes listed, namely respiration and ethylene production, have gained priority in attempts to develop an explanation of the mechanism of fruit ripening. Further characterization of other changes occurring in climacteric and non-climacteric fruit is given later in this chapter (see 'Chemical changes during maturation').

Physiology of respiration

A major metabolic process taking place in harvested produce or in any living plant product is respiration. Respiration can be described as the oxidative breakdown of the more complex materials normally present in cells, such as starch, sugars and organic acids, into simpler molecules, such as carbon dioxide and water, with the concurrent production of energy and other molecules which can be used by the cell for synthetic reactions. Respiration can occur in the presence of oxygen (aerobic respiration) or in the absence of oxygen (anaerobic respiration, sometimes called fermentation).

Respiration rate of produce is an excellent indicator of metabolic activity of the tissue and thus is a useful guide to the potential storage life of the produce. If the respiration rate of a fruit or vegetable is measured—as either oxygen consumed or carbon dioxide evolved—during the course of its development, maturation, ripening and senescent periods, a characteristic respiratory pattern is obtained. Respiration rate per unit weight is highest for the immature fruit or vegetable

Figure 9   Growth and respiration patterns of fruit during development. (From Biale, J.B. 'Growth, maturation, and senescence in fruits.' Science 146: 880–8; 1964. With permission.)

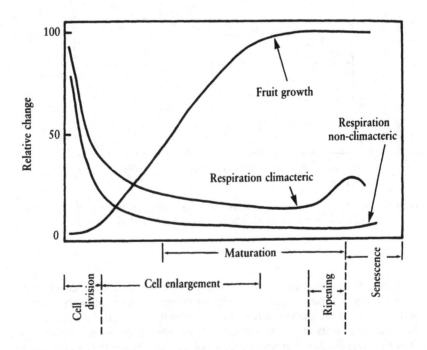

and then steadily declines with age (Figure 9). A significant group of fruits that includes tomato, mango, banana and apple, shows a variation from the described respiratory pattern in that they undergo a pronounced increase in respiration coincident with ripening (Figure 9). Such an increase in respiration is known as a respiratory climacteric, and this group of fruits is known as the climacteric class of fruits. The intensity and duration of the respiratory climacteric, first described in 1925 for the apple, varies widely amongst fruit species as depicted in Figure 10. The commencement of the respiratory climacteric coincides approximately with the attainment of maximum fruit size (Figure 9), and it is during the clim-

Figure 10 Respiratory patterns of some climacteric fruits. (From Biale, J.B. 'Postharvest physiology and biochemistry of fruits.' Ann. Rev. Plant Physiol. 1: 183–206; 1950. With permission.)

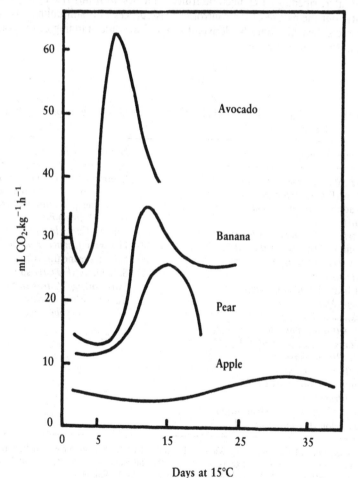

acteric that all the other changes characteristic of ripening occur. The respiratory climacteric, as well as the complete ripening process, may proceed while the fruit is either attached to or detached from the plant. Those fruits, such as citrus, pineapple and strawberry, that do not exhibit a respiratory climacteric, are known as the non-climacteric class of fruit. Non-climacteric fruit exhibit most of the ripening changes, although these usually occur more slowly than those of the climacteric fruits. Table 8 lists some common climacteric and non-climacteric fruits. All vegetables can also be considered to have a non-climacteric type of respiratory pattern.

The division of fruit into two classes on the basis of their respiratory pattern is an arbitrary classification but has served to stimulate considerable research to discover the biochemical control of the respiratory climacteric. As will be described later, ripening of climacteric fruits coincides with this rise in respiratory activity, but the nature of the control of these processes is still unknown. Large commercial benefits could be derived from a clear understanding of the control mechanism of fruit ripening.

Table 8   Classification of some edible fruits according to their respiratory behaviour during ripening[1]

Climacteric fruits	Non-climacteric fruits
Apple (*Malus domestica*)	Cherry: sweet (*Prunus avium*)
Apricot (*Prunus armeniaca*)	sour (*Prunus cerasus*)
Avocado (*Persea americana*)	Cucumber (*Cucumis sativus*)
Banana (*Musa* sp.)	Grape (*Vitis vinifera*)
Blueberry (*Vaccinium corymbosum*)	Lemon (*Citrus limon*)
Cherimoya (*Annona cherimola*)	Pineapple (*Ananas comosus*)
Feijoa (*Feijoa sellowiana*)	Satsuma mandarin (*Citrus unshu*)
Fig (*Ficus carica*)	Strawberry (*Fragaria* sp.)
Kiwi fruit (*Actinidia deliciosa*)	Sweet orange (*Citrus sinensis*)
Mango (*Mangifera indica*)	Tamarillo (tree tomato)
Muskmelon (*Cucumis melo*)	(*Cyphomandra betacea*)
Papaya (*Carica papaya*)	
Passionfruit (*Passiflora edulis*)	
Peach (*Prunus persica*)	
Pear (*Pyrus communis*)	
Persimmon (*Diospyros kaki*)	
Plum (*Prunus* sp.)	
Tomato (*Lycopersicon esculentum*)	
Watermelon (*Citrullus lanatus*)	

[1] Amended from McGlasson, W.B.; Wade, N.L.; and Adato, I. Phytohormones and fruit ripening. Letham, D.S.; Goodwin, P.B.; Higgins, T.J.V. eds. Phytohormones and related compounds: a comprehensive treatise. Vol. 2. Amsterdam: Elsevier; 1978: 447–93. With permission.

Effect of ethylene

Climacteric and non-climacteric fruits may be further differentiated by their response to applied ethylene and by their pattern of ethylene production during ripening. It has been clearly established that all fruit produces minute quantities of ethylene during development. However, coincident with ripening, climacteric fruits produce much larger amounts of ethylene than non-climacteric fruits. This difference between the two classes of fruit is further exemplified by the internal ethylene concentration found at several stages of development and ripening (Table 9). The internal ethylene concentration of climacteric fruits varies widely, but that of non-climacteric fruits changes little during development and ripening. Ethylene, applied at a concentration as low as 0.1–1.0 microlitres per litre for one day, is normally sufficient to hasten full ripening of climacteric fruits (Figure 11), but the magnitude of the climacteric is relatively independent of the concentration of applied ethylene. In contrast, applied ethylene merely increases the respiration of non-climacteric fruits, the magnitude of the increase being dependent on the concentration of ethylene (Figure 11). Moreover, the rise in respiration in response to ethylene may occur more than once in non-climacteric fruits in contrast to the single respiration increase in climacteric fruits.

The significance of ethylene for fruit ripening was established during the early part of this century when heaters burning kerosene were used to degreen—or colour yellow—California lemons. Denny, in 1924, found that while warmth

Table 9 Internal ethylene concentrations measured in several climacteric and non-climacteric fruits[1]

Fruits	Ethylene (μL/L)
Climacteric	
Apple	25–2500
Pear	80
Peach	0.9–20.7
Nectarine	3.6–602
Avocado	28.9–74.2
Banana	0.05–2.1
Mango	0.04–3.0
Passionfruit	466–530
Plum	0.14–0.23
Tomato	3.6–29.8
Non-climacteric	
Lemon	0.11–0.17
Lime	0.30–1.96
Orange	0.13–0.32
Pineapple	0.16–0.40

[1] From Burg, S.P., Burg, E.A. The role of ethylene in fruit ripening. Plant Physiol. 37: 179–89: 1962. With permission.

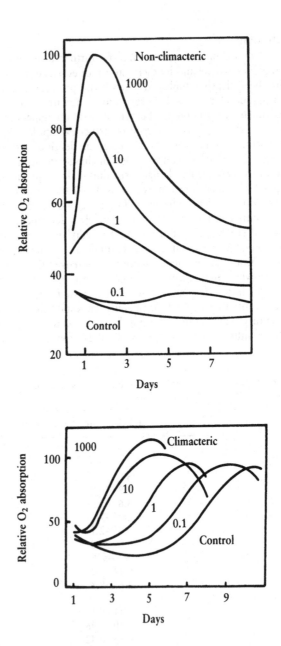

Figure 11 Effect of applied ethylene on respiration of climacteric and non-climacteric fruits. (From Biale, J.B. 'Growth, maturation, and senescence in fruits.' Science 146: 880–8; 1964. With permission.)

was needed, the real cause of degreening was ethylene, and many workers soon demonstrated that ethylene could hasten the ripening of many fruits. Ethylene was regarded as an external agent that could promote the ripening of fruit, but in 1934 Gane found that fruit, and other plant tissues, produced extremely small quantities of ethylene. This finding was achieved despite the use of insensitive and time-consuming methods of analysis, which generally relied on trapping the ethylene with mercuric perchlorate, releasing the trapped ethylene and measuring it by manometry.

Research into the involvement of ethylene in fruit ripening was greatly stimulated by the development of sensitive, gas chromatographic techniques for the measurement of low levels of ethylene, so that it is now possible to measure quantitatively as little as 0.001 microlitre per litre in a 1 millilitre gas sample. This obviates the need for collecting or absorbing ethylene over a lengthy period of time.

Role of ethylene
Ethylene biosynthesis
Ethylene has been shown to be produced from methionine via a pathway that includes the intermediates S-adenosyl-methionine (SAM) and 1-aminocyclopropane-1-carboxylic acid (ACC). The conversion of SAM to ACC by the enzyme ACC synthase is thought to be the rate limiting step in the biosynthesis of ethylene. However, addition of ACC to preclimacteric (unripe) fruit generally results in only a small increase in ethylene evolution showing that another enzyme, the ethylene-forming enzyme (EFE), is required to convert ACC to ethylene. EFE has not been identified but it is known to be labile and is thought to be membrane-bound. Factors that affect the activity of ACC synthase include fruit ripening, senescence, auxin, physical injuries, and chilling injury. This enzyme is believed to be a pyridoxal enzyme because it requires pyridoxal phosphate for maximal activity and is strongly inhibited by (aminooxy) acetic acid (AOA) and L-2-amino-4-(2-aminoethoxy)-*trans*-3-butenoic acid (AVG) which are known inhibitors of pyridoxal phosphate-dependent enzymes. EFE is inhibited by anaerobiosis, temperatures above 35°C and cobalt ions. Small amounts of ethylene can also be formed in plant tissues from the oxidation of lipids involving a free-radical mechanism.
Mode of action
Ethylene is a plant hormone which probably acts in concert with other plant hormones (auxins, gibberellins, kinins and abscisic acid) to exercise control over the fruit ripening process. Most is known about the relation of ethylene to fruit ripening because the availability of a sensitive, physical method for measurement of ethylene has enabled detailed studies of this relationship. The relationship of the other plant hormones to ripening is as yet not clearly defined.

It has been proposed that two systems exist for the regulation of ethylene biosynthesis: system 1 is initiated or perhaps controlled by an unknown factor, probably involved in the regulation of senescence. System 1 then triggers system 2 which is responsible, during ripening of climacteric fruits, for the production of the large amounts of ethylene that are necessary for the full integration of ripening. Non-climacteric fruits do not have an active system 2, and treatment of climacteric fruits with ethylene circumvents system 1.

As in the case of other plant hormones, ethylene is thought to bind to specific receptor(s) to form a complex which then triggers ripening. Ethylene action can be affected by altering the amount of receptors(s) or by interfering with the binding of ethylene to its receptor. Detailed studies of the structural requirements for biological activity of ethylene receptors led to the proposal that binding takes place reversibly at a site containing a metal, possibly copper. From kinetic studies on the responses of plant tissue to added ethylene it has been proposed that the affinity of the receptor for ethylene is increased by the presence of oxygen and decreased by carbon dioxide. The occurrence of a metal-containing receptor has as yet not been confirmed but the idea is supported by studies with silver ion. Treatment of fruit, flowers and other tissues with silver ion has been shown to inhibit the action of ethylene. The need for specific structural requirements for ethylene action has been demonstrated by treating tissues with analogues and antagonists of ethylene. The gaseous cyclic olefine, 2,5-norbornadiene has been shown to be a highly effective inhibitor of ethylene action.

The pattern of changes in ethylene production rates and the internal concentrations of ethylene in relation to the onset of ripening have been observed in several climacteric fruits. In one type of fruit ethylene concentration rises before the onset of ripening determined as the initial respiratory increase, eg banana, tomato, and honey dew melon. In the second type, ethylene does not rise before the increase in respiration, eg apple, avocado, and mango. In honey dew melon the internal ethylene concentration rises from the preclimacteric level of 0.04 microlitre/litre to 3.0 microlitres/litre at which concentration the fruit commences

Table 10   Effect of maturity on the time to ripen for tomato.[1]
Time to ripen was determined between anthesis and the first detectable red colour (first colour stage). Fruit were treated continuously with 1000 μL/L ethylene

Maturity at harvest (Days after anthesis)	Days to ripen	
	Treated with ethylene	Control
17	11	—[2]
25	6	—
31	5	15
35	4	9
42	1	3

[1] Lyons. J.M.; Pratt, H.K. Effect of stage of maturity and ethylene treatment on respiration and ripening of tomato fruits. Proc. Am. Soc. Hortic. Sci. 84: 491–500: 1964. With permission.
[2] Had failed to ripen when experiment was terminated.

to ripen. The low concentrations of ethylene present in unripe fruits and the evident involvement of system 2 ethylene in ripening indicate that treatments which prevent ethylene from reaching a triggering concentration should delay ripening. Ripening has been delayed in green banana fruit for up to 180 days at 20°C when the fruit were ventilated continuously with an atmospheres of 5 per cent carbon dioxide, 3 per cent oxygen, and 92 per cent nitrogen or when bulk samples of fruit were stored in modified atmospheres in the presence of an ethylene absorbent such as potassium permanganate which oxidises this hydrocarbon.

It is well known that many fruits, as they develop and mature, become more sensitive to ethylene. For some time after anthesis (flowering) young fruit can have high rates of ethylene production. Early in the life of fruit the concentration of applied ethylene required to initiate ripening is high, and the length of time to ripen is prolonged but decreases as the fruit matures (Table 10). The tomato is an extreme case of tolerance to ethylene. Banana and melons, in contrast, can be readily ripened with ethylene even when immature. Nothing is known about the factor(s) that control the sensitivity of the tissue to ethylene.

Ripening has long been considered to be a process of senescence and to be due to a breaking down of the cellular integrity of the tissue; some ultrastructural and biochemical evidence supports this view. There is now considerable evidence for ripening being a programmed phase in the differentiation of plant tissue, with altered nucleic acid and protein synthesis occurring at the commencement of the respiratory climacteric. Both views fit with the known degradative and synthetic capacities of fruit during ripening. In view of the ample evidence of the ability of ethylene to initiate biochemical and physiological events it seems likely that ethylene action can be regulated at the level of gene expression.

BIOCHEMISTRY OF RESPIRATION

All living organisms require a continuous supply of energy. This energy enables the organism to carry out the necessary metabolic reactions to maintain cellular organization, to transport metabolites around the tissue and to maintain membrane permeability.

Aerobic metabolism

Most of the energy required by fruit and vegetables is supplied by aerobic respiration, which involves the oxidative breakdown of certain organic substances stored in the tissues. The normal substrate for respiration is glucose, and, if it is completely oxidized, the overall reaction is:

$$C_6H_{12}O_6 + 6O_2 \rightarrow 6CO_2 + 6H_2O + \text{energy}$$

Respiration is essentially the reverse of photosynthesis by which energy derived from the sun is stored as chemical energy, mainly in carbohydrates containing glucose. Full utilization of glucose involves two main reaction sequences:

1. glucose → pyruvate; by the Embden-Meyerhof-Parnas (EMP) pathway, located in the cytoplasm;

2. pyruvate → carbon dioxide; by the TCA cycle, the enzymes of which are located in mitochondria.

Free glucose is conventionally the compound involved in the initial oxidative step, but it is not the storage form of carbohydrate in the plant. Starch, a polymer of glucose, is often the main carbohydrate, and it must be degraded first to glucose by amylases and maltase. Other commodities have a high sucrose content which can be hydrolyzed to glucose and fructose by the enzyme invertase. Interconversion of sucrose and starch is also possible in many plant tissues. A generalized scheme for the initial conversions of the storage carbohydrates is shown in Figure 12.

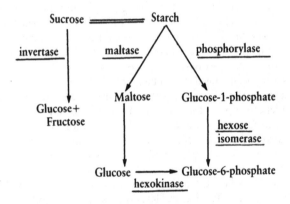

Figure 12 Degradation of carbohydrate storage reserves.

The EMP sequence

A simplified view of the EMP sequence is shown in Figure 13. The overall reaction system can be balanced as:

$$\text{glucose} + 2\text{ADP} + 2\text{P}_i + 2\text{NAD} \rightarrow 2\text{pyruvate} + 2\text{ATP} + 2\text{NADH}_2 + 2\text{H}_2\text{O}$$

The energy liberated from the reaction system is trapped and stored in adenosine triphosphate (ATP) and reduced nicotinamide adenine dinucleotide (NADH_2), where each molecule of NADH_2 gives 3ATP. The total energy liberated by the conversion of glucose to pyruvate is, therefore, equivalent to 8 ATP. The energy is subsequently made available to the plant by breaking a phosphate bond in the reverse reaction:

$$\text{ATP} \rightarrow \text{ADP} + \text{P}_i + \text{energy}$$

An alternate method for the conversion of glucose to phosphoglyceraldehyde involves the formation and interconversion of phosphorylated C_5 sugars, such as ribulose, ribose and xylulose, but this cycle is normally considered to be less important than the EMP sequence.

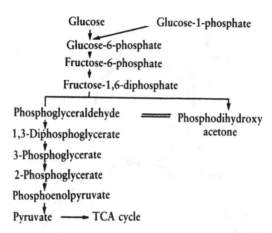

Figure 13 EMP pathway.

The TCA cycle

A simplified view of the full cycle is shown in Figure 14. The overall reaction system is:

$$\text{pyruvate} + 3O_2 + 15ADP + 15P_i \rightarrow 3CO_2 + 2H_2O + 15ATP$$

The energy of the original glucose molecule (giving 2 × pyruvate) liberated from the TCA cycle is 30ATP (compared with 8ATP from the EMP sequence). The carbon dioxide produced in respiration is derived from the TCA cycle under aerobic conditions and involves a consumption of oxygen. The rate of respiration can, therefore, be measured by the amount of carbon dioxide produced or oxygen consumed.

The total chemical energy liberated during the oxidation of 1 mole of glucose is approximately 1.6 megajoules. About 90 per cent of this energy is preserved within the plant system, and the remainder is lost as heat. Respiration is, therefore, an efficient converter of energy compared with man-made devices for energy conversion, for example, in the petrol engine over 50 per cent of the liberated energy is lost as heat.

Respiration quotient (RQ)

The vacuoles of many fruits and vegetables have large reserves of organic acids that can be mobilized for use in the mitochondria as oxidizable substrates in the TCA cycle. The complete oxidation of malate

$$C_4H_6O_5 + 3O_2 \rightarrow 4CO_2 + 3H_2O$$

generates more carbon dioxide than the amount of oxygen consumed, whereas oxidation of glucose generates an equal amount of carbon dioxide for the oxygen consumed. This relationship becomes important when measuring respiration by

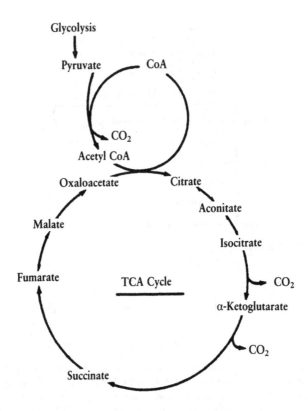

Figure 14 TCA cycle.

gas exchange, in which the carbon dioxide evolved and/or oxygen consumed is measured, that is, it is possible to record different values for respiration depending on which gas is monitored. Ideally both gases should be measured simultaneously.

The concept of respiration quotient has been developed to quantify this variation, where:

$$RQ = CO_2 \text{ produced (mL)}/O_2 \text{ consumed (mL)}$$

For the complete oxidation of glucose, RQ = 1.0, whereas for malate RQ = 1.3. An alternative substrate could be long-chain fatty acids, for example, stearic acid:

$$C_{18}H_{36}O_2 + 26O_2 \rightarrow 18CO_2 + 18H_2O$$

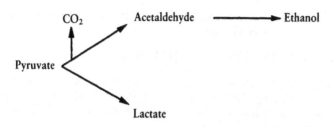

Figure 15 Pathways of anaerobic metabolism.

These fatty acids have much less oxygen per carbon atom than sugars and, therefore, require a greater oxygen consumption for the production of CO_2. RQ equals 0.7 for the above reaction.

Measurement of RQ itself can give some guide to the type of substrate that is being respired: a low RQ suggests some fat metabolism and a high RQ suggests organic acids. Changes in RQ during growth and storage can also indicate a change in the type of substrate that is being metabolized.

Anaerobic metabolism (fermentation)

These respiratory pathways utilize oxygen and are the preferred pathways in fruit and vegetables. The normal atmosphere is rich in oxygen, so the amount of oxygen available in the tissue is unlimited. Under various storage conditions the amount of oxygen in the atmosphere may be limited and insufficient to maintain full aerobic metabolism. Under these conditions the tissue can initiate anaerobic respiration, by which glucose is converted to pyruvate by the EMP pathway. But pyruvate is then metabolized into either lactic acid or acetaldehyde and ethanol in a process termed fermentation (Figure 15). The oxygen concentration at which anaerobic respiration commences, varies between tissues and is known as the extinction point. The oxygen concentration at this point depends on several factors, such as species, cultivar, maturity and temperature. Anaerobic respiration produces much less energy per mole of glucose than aerobic pathways, but it does allow some energy to be made available to the tissue under adverse conditions. A high RQ is generally indicative of fermentation reactions.

Metabolites for synthetic reactions

The respiratory pathways are not only used for the production of energy for the tissue. Carbon skeletons are required for many synthetic reactions in the cell, and these skeletons can be removed at several points. For example, α-ketoglutarate may be converted to the amino acid glutamate from which several other amino acids may be produced for protein synthesis; succinate may be diverted into the synthesis of various heme pigments including chlorophyll. The loss of α-ketoglutarate and succinate from the TCA cycle for synthetic reactions would eventually lead to the stopping of the cycle. Therefore, C4 acids are fed into the cycle; they are produced principally by the fixation of carbon dioxide into phosphoenol-pyruvate to give oxaloacetate. Alternatively, vacuolar reserves of, for example, malate may be utilized.

CHEMICAL CHANGES DURING MATURATION

At some stage during the growth and development of fruit and vegetables, the produce is recognised by the consumer as having attained optimum eating condition. This desirable quality is not associated with any universal change, but is attained in various ways in different tissues.

Fruit

Climacteric fruits generally reach the fully ripe stage after the respiratory climacteric. However, it is the other events initiated by ethylene that the consumer associates with ripening.

Colour

Colour is the most obvious change that occurs in many fruits and is often the major criterion used by consumers to determine whether the fruit is ripe or unripe. The most common change is the loss of green colour. With a few exceptions, for example, the avocado and Granny Smith apple, climacteric fruits show rapid loss of green colour on ripening. Many non-climacteric fruits also exhibit a marked loss of green colour with attainment of optimum eating quality, for example, citrus fruit in temperate climates (but not in tropical climates). The green colour is due to the presence of chlorophyll which is a magnesium-organic complex. The loss of green colour is due to degradation of the chlorophyll structure. The principal agents responsible for this degradation are pH changes (mainly due to leakage of organic acids from the vacuole), oxidative systems and chlorophyllases (Figure 16). Loss of colour depends on one or all of these factors acting in sequence to destroy the chlorophyll structure.

The disappearance of chlorophyll is associated with the synthesis and/or revelation of pigments ranging from yellow to red. Many of these pigments are carotenoids, which are unsaturated hydrocarbons with generally forty carbon atoms and which may have one or more oxygen functions in the molecule. Carotenoids are stable compounds and remain intact in the tissue even when extensive senescence has occurred. Carotenoids may be synthesized during the development stages on the plant, but they are masked by the presence of chlorophyll. Following the degradation of chlorophyll, the carotenoid pigments become visible; with other tissues, carotenoid synthesis occurs concurrently with chlorophyll degradation. Banana peel is an example of the former system and tomato of the latter.

Anthocyanins provide many of the red-purple colours of fruit and vegetables. Anthocyanins are water soluble so they are found mainly in the cell vacuoles of fruit and vegetables, often in the epidermal layers. They produce strong colours, which often mask carotenoids and chlorophyll.

Carbohydrates

The largest quantitative change associated with ripening is usually the breakdown of carbohydrate polymers, particularly frequent near total conversion of starch to

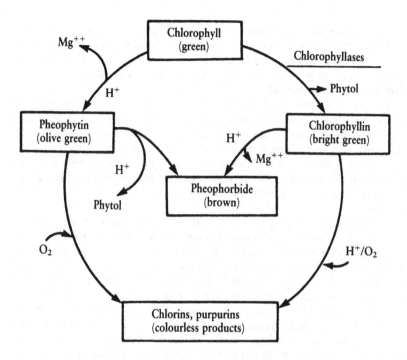

Figure 16   Some pathways for the degradation of chlorophyll.

sugars. This has the dual effect of altering the taste and texture of the produce. The increase in sugar renders the fruit much sweeter and, therefore, more acceptable. Even with non-climacteric fruits, the accumulation of sugar is associated with the development of optimum eating quality, although the sugar may be derived from the sap imported into the fruit rather than from the breakdown of starch reserves of the fruit.

The breakdown of polymeric carbohydrates, especially pectic substances and hemicelluloses, weakens cell walls and the cohesive forces binding cells together. In the initial stages, the texture becomes more palatable, but eventually the plant structures disintegrate. Protopectin is the insoluble parent form of pectic substances. In addition to being a large polymer, it is cross-linked to other polymer chains with calcium (Ca) bridges and is bound to other sugars and phosphate derivatives to form an extremely large polymer. During ripening and maturation, protopectin is gradually broken down to lower molecular weight fractions, which are more soluble in water. The rate of degradation of pectic substances is directly correlated with the rate of softening of fruit.

Organic acids
Usually organic acids decline during ripening as they are respired or converted to sugars. Acids can be considered as a reserve source of energy to the fruit and

would, therefore, be expected to decline during the greater metabolic activity that occurs on ripening. There are exceptions, such as banana and pineapple, where the highest levels are attained at the full ripe stage, but the levels in these fruits are not high at any stage of development compared to other produce.

Nitrogenous compounds

Proteins and free amino acids are minor constituents of fruit and, as far as is known, have no role in determining eating quality. Changes in nitrogenous constituents do, however, indicate variations in metabolic activity during different growth phases. During the climacteric phase of many fruits, there is a decrease in free amino acids which often reflects an increase in protein synthesis. During senescence, the level of free amino acids increases reflecting a breakdown of enzymes and decreased metabolic activity.

Aroma

Aroma plays an important part in the development of optimal eating quality in most fruit. It is due to the synthesis of many volatile organic compounds (often known merely as volatiles) during the ripening phase. The total amount of carbon involved in the synthesis of volatiles is less than 1 per cent of that expelled as carbon dioxide. The major volatile formed is ethylene, which accounts for about 50–75 per cent of the total carbon in the volatiles; ethylene does not contribute to typical fruit aromas. The amount of aroma compounds is therefore extremely small. Chapter 2 discussed the nature of the compounds formed. Non-climacteric fruits also produce volatiles during the development of optimum eating quality. These fruits do not synthesize compounds that are as aromatic as those in climacteric fruit; nevertheless, the volatiles produced are still important in consumer appreciation.

Vegetables

Vegetables generally show no sudden increase in metabolic activity that parallels the onset of the climacteric in fruit, unless sprouting or regrowth is initiated. The process of germination is sometimes deliberately applied to some seeds, for example, mung bean, and the sprouted product is the marketed vegetable. Apart from obvious anatomical changes during sprouting, considerable compositional changes occur. The sugar level increases markedly as the result of the rapid conversion of fats or starch. From a nutritional view, the increase in vitamin C in sprouted seeds can be valuable in diets with marginal vitamin C intakes. Vegetables can be divided into three main groups: seeds and pods; bulbs, roots and tubers; flowers, buds, stems and leaves. Some fruits are also consumed as vegetables; they may be either ripe (tomato, egg-plant) or immature (zucchini, cucumber, okra).

Seeds and pods, if harvested fully mature, as is the practice with cereals, have low metabolic rates because of their low water content. In contrast, all seeds consumed as fresh vegetables, for example, legumes and sweet corn, have high levels of metabolic activity, because they are harvested at an immature stage, often with the inclusion of non-seed material, for example, bean pod (pericarp). Eating quality is determined by flavour and texture and not by physiological age.

Generally the seeds are sweeter and more tender at an immature stage. With advancing maturity, the sugars are converted to starch with the resultant loss of sweetness, the water content decreases and the amount of fibrous material increases. Seeds for consumption as fresh produce are harvested when the water content is about 70 per cent; in contrast, dormant seeds are harvested at less than 15 per cent water.

Bulbs, roots and tubers are storage organs that contain food reserves that are required when growth of the plant is resumed (they are often held for the purpose of propagation). When harvested, their metabolic rate is low and, under appropriate storage conditions, their dormancy can be prolonged.

Edible flowers, buds, stems and leaves vary greatly in metabolic activity and hence in rate of deterioration. Stems and leaves often senesce rapidly and so lose their attractiveness and nutritive value. Texture often becomes the dominant characteristic that determines both the harvest date and quality. The natural flavour is often of less importance than texture, as many of these vegetables are cooked and salt or spice added.

FURTHER READING

Brady, C.J. Fruit ripening. Annu. Rev. Plant Physiol. 38: 155–78; 1987.

Bronk, J.R. Chemical biology—an introduction to biochemistry. New York: Macmillan Inc.: 1973.

Burton, W.G., Postharvest physiology of food crops. New York: Longman: 1982.

Conn, E.E.; Stumpf, P.K. Outlines of biochemistry. 3d ed. New York: Wiley: 1972.

Coombe, B.G. The development of fleshy fruits. Annu. Rev. Plant Physiol. 27: 507–28; 1976.

Duckworth, R.B. Fruit and vegetables. Oxford: Pergamon Press: 1966.

Friend, J.; Rhodes, M.J.C., eds. Recent advances in the biochemistry of fruits and vegetables. London: Academic Press; 1981.

Fuchs, Y.; Chalutz, E. Ethylene: biochemical, physiological and applied aspects. The Hague: Martinus Nijhoff; 1984.

Goodwin, T.W., ed. Chemistry and biochemistry of plant pigments. 2d ed. Vol. I. London: Academic Press; 1976.

Hulme, A.C., ed. The biochemistry of fruits and their products. Vols 1 and 2. London: Academic Press; 1970, 1971.

Hultin, H.O.; Milner, M., eds. Postharvest biology and biotechnology. Westport, CT: Food & Nutrition Press; 1978.

Jones, C.W. Biological energy conservation. Outline studies in biology. London: Chapman and Hall; 1976.

Kader, A.A.; Morris, L.L.; Cantwell, M. Postharvest handling and physiology of horticultural crops—a selected list of references. 3d ed. Davis, CA: University of California; 1983. Vegetable Crops Series 169.

Lallu, N. Post-harvest aspects of Asian pears. I. Maturity. Asian pear series. Publication No. 7. Wellington: N.Z. Apple and Pear Marketing Board, Technical Bull. No. 8; 1985.

Leopold, A.C.; Kriedemann, P.E. Plant growth and development. 2d ed. New York: McGraw-Hill; 1975.

Leiberman, M. Post-harvest physiology and crop preservation. New York: Plenum; 1983.

Loewy, A.G.; Siekevitz. P. Cell structure and function. New York: Holt, Rinehart and Winston, 1969.

McGlasson, W.B. Ethylene and fruit ripening. HortScience 20: 51–54; 1985.

McGlasson, W.B.; Wade, N.L.; Adato, I. Phytohormones and fruit ripening. Letham, D.S.; Goodwin, P.B.; Higgins, T.J.V. eds. Phytohormones and related compounds: a comprehensive treatise. Vol. 2. Amsterdam: Elsevier; 1978: 447–93.

Pantastico, E.B., ed. Postharvest physiology, handling and utilization of tropical and subtropical fruits and vegetables. Westport CT: AVI: 1975.

Pratt, H.K.; Goeschl, J.D. Physiological roles of ethylene in plants. Annu. Rev. Plant Physiol. 20: 541–84; 1969.

Roberts, J.; Tucker, G.A., eds. Ethylene and plant development. Oxford: Butterworths; 1985.

Sacher, J.A. Senescence and postharvest physiology. Annu. Rev. Plant Physiol. 24: 197–224; 1973.

Salunkhe, D.K.; Desai, B.B. Postharvest biotechnology of fruits. Vols 1 and 2. Boca Raton, FL: CRC Press: 1984.

Salunkhe, D.K. Postharvest biotechnology of vegetables. Vols 1 and 2. Boca Raton, FL: CRC Press: 1984.

Subramanyam, H.; Krishnamurthy, S.; Parpia, H.A.B. Physiology and biochemistry of mango fruit. Adv. Food Res. 21: 223–305; 1975.

Wardowski, W.F.; Nagy, S.; Grierson, W. Fresh citrus fruits. Westport, CT: AVI; 1986.

Weichmann, J., ed. Postharvest physiology of vegetables. New York: Marcel Dekker; 1987.

Yang, S.F. Biosynthesis and action of ethylene. HortScience 20: 41–45: 1985.

Yang, S.F.; Hoffman, N.E. Ethylene biosynthesis and its regulation in higher plants. Annu. Rev. Plant Physiol. 35: 155–89; 1984.

4
Effects of temperature

Respiration in fruit and vegetables involves many enzymic reactions (Chapter 3). The rate of these reactions, within the physiological temperature range, increases exponentially with increase in temperature and may be described mathematically by use of the temperature quotient (Q_{10}). Van't Hoff, the Dutch chemist, showed that the rate of a chemical reaction approximately doubles for each 10°C rise in temperature:

$$Q_{10} = (R_2/R_1)^{10(t_2 - t_1)} = \text{constant, about 2}$$

where t_2 and t_1 are any temperatures (degrees Celsius) and R_2 and R_1 are the respective rates. From this formula, either the Q_{10} or an unknown rate for any temperature difference may be calculated.

For many biological processes, however, the Q_{10} does not remain constant over the physiological range, that is, Q_{10} is a function of temperature. Q_{10} values are generally highest between 1°C and 10°C and can be as high as seven, but at temperatures above 10°C, Q_{10} values generally fall between two and three.

PHYSIOLOGICAL PESPONSES

The activity of enzymes in fruit and vegetables declines at temperatures above 30°C, but the temperature at which specific enzymes become inactive varies. Many are still active at 35°C, but most are inactivated at 40°C. Continuous exposure of some climacteric fruits to a temperature of about 30°C causes the flesh to ripen, but the fruit fails to colour normally, for example, the peel of Cavendish banana (cultivars Valery and Williams) remains green, and lycopene (red pigment) accumulation in tomato is inhibited. When produce is held above 35°C, metabolism becomes abnormal and results in a breakdown of membrane integrity and structure, with disruption of cellular organization and rapid deterioration of the produce. The changes are often characterized by a general loss

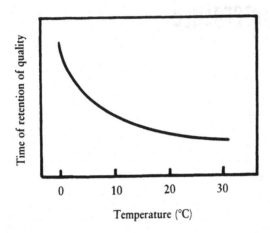

Figure 17   Effect of temperature on retention of quality.

of pigments, and the tissues may develop a watery or translucent appearance. Such a condition in banana and tomato is referred to as 'boiled'.

The lower limit for normal metabolism is the freezing point of the tissues, which is usually between 0°C and −2°C. Once the tissue is frozen the interchange of metabolites among the various cellular components is seriously hampered. Much of the water freezes outside the cell, resulting in permanent desiccation of cells. The expansion of the water upon freezing also causes considerable damage to the cell. Upon thawing, the tissue invariably fails to resume normal metabolism and to regain normal texture. Ideally the greatest reduction in respiration and general metabolism, and thus longest life, will be obtained if the produce is held just above its freezing point.

Reduction in temperature will give some reduction in the rate of change of any given parameter whether it be respiration, change in texture or loss of vitamin C (Figure 17). However, the effect of reducing the temperature is not uniform for all physiological factors. Only a small improvement in storage life (i.e. the time that produce can be held in an acceptable condition after harvest) is achieved by small reductions in temperature at the upper end of the temperature range. Much larger improvements are obtained by small reductions at the lower temperatures (Figure 18), where even a change in temperature of 1°C can have a significant effect. Cooling below 10°C, except for brief periods, does not benefit produce that is sensitive to chilling injuries (Chapter 7). Additional benefits can be gained by the greatly diminished rate of microbial growth at low temperatures. If the temperature is low enough, many fungal spores will not germinate (Chapter 9).

Lowering the temperature of non-climacteric produce merely lowers the rate of deterioration, whereas in climacteric fruits low temperatures can also be used to delay the onset of ripening. The effect of decreased temperature on ripening

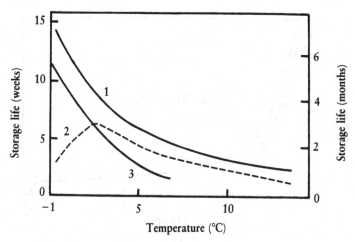

Figure 18 Effect of temperature on storage life of apple and pear. 1. Delicious apple (months); 2. apple cultivars susceptible to low temperature disorders (months); 3. Williams Bon Chretien (Bartlett) pear (weeks).

follows an exponential relationship similar to that shown in Figure 18. Lowering the temperature decreases not only production of ethylene, but also the rate of response of the tissues to ethylene, so that at lower temperatures longer exposure to a given concentration of ethylene is required to initiate ripening. Normal ripening occurs only within a particular range of temperatures (commonly 10–30°C) although some fruits, for example, some pear cultivars (Figure 19), will ripen slowly and satisfactorily at temperatures below 10°C. The best quality in ripe fruit generally develops at about 20°C, a temperature which is considered to be the optimum for the ripening of most fruits.

Provided that the fruit is not sensitive to chilling, maximum storage life can be achieved at temperatures below the ripening range. For example, Williams Bon Chretien (WBC) (or Bartlett) pears will not ripen at temperatures below about 12°C, but maximum storage life is obtained by storage at −1°C and removal to temperatures greater than 12°C when ripening is desired (Figure 18, 19). If the pears, however, are held too long at low, non-ripening temperatures, they will fail to ripen normally after removal to ripening temperatures, probably due to inactivation or destruction of the enzymes necessary for ripening. In addition to abnormal ripening, storage at low temperatures can lead to other metabolic abnormalities which lead to localized cell collapse (Chapter 7).

Sugar–starch balance

Storage of some vegetables, for example, potato, sweet potato, peas and sweet corn, at low temperatures alters the starch-sugar balance in the produce. At any temperature, starch and sugar are in dynamic equilibrium, and some sugar is degraded to carbon dioxide during respiration:

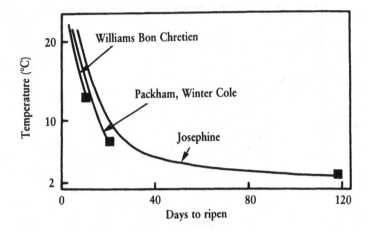

Figure 19 Effect of temperature on rate of ripening of four pear cultivars. ■ Signifies will not ripen normally below this temperature.

$$\text{starch} \rightleftharpoons \text{sugar} \rightarrow CO_2$$

At ambient temperatures, the starch-sugar balance in potato and sweet potato is heavily biased towards the accumulation of starch. When these vegetables are stored below a critical temperature, the rate of respiration and the conversion of sugar to starch decreases, and sugar accumulates in the tissues. The critical temperature at which accumulation of sugar commences depends on the commodity, for example, about 10°C for potato and about 15°C for sweet potato. The accumulation of sugar is undesirable in many starchy vegetables. Potato with a high sugar content has a poor texture and a sweet taste when boiled: when fried, excessive browning develops due to caramelization and reactions between amino acids and sugars (Maillard reaction). The accumulation of sugar in potato stored at low temperatures can be largely reversed by raising the storage temperature to 10°C or above. Although it is generally accepted that the sugar level returns to nearly normal during one week at 15–20°C, experience has shown that the decrease in reducing sugar level may occur at a slower rate.

In other vegetables, such as sweet corn and peas, a high sugar content is desired. These vegetables are harvested immature when the sugar content is highest, and storage at low temperatures is desirable to retard conversion of sugar to starch.

STORAGE LIFE

There is no one ideal temperature for the storage of all fruit and vegetables, because their responses to reduced temperatures vary widely. The importance of factors such as mould growth and chilling injury must be taken into account, as

Table 11 Typical respiration rates of several fruits and vegetables[1]

Fruit	Respiration rate at 15°C (mL $CO_2kg^{-1}h^{-1}$)*	Vegetable	Respiration rate at 15°C (mL $CO_2kg^{-1}h^{-1}$)*
Apple	25	Bean	250
Banana, green	45	Cabbage	32
Banana, ripe	200	Carrot	45
Grape	16	Lettuce	200
Lemon	20	Pea	260
Orange	20	Potato	8
Peach	50		
Pear	70		
Strawberry	75		

* Millilitres of carbon dioxide per kilogram per hour.
[1] Values adapted from American Society of Heating. Refrigerating and Air-Conditioning Engineers. (1974).

Table 12 Storage life of fresh fruits and vegetables[1]

Produce	Time at optimum temperature (weeks)		
	−1–4°C	5–9°C	10°C
Fruit			
Very perishable (0–4 weeks)			
Apricot	2		
Banana, ripe			1–2
Banana, green			1–2[2]
Berry fruits	1–2		
Cherry	1–4		
Fig	2–3		
Loquat	1–2		
Mango		2–3	
Strawberry	1–5 days		
Watermelon		2–3	
Perishable (4–8 weeks)			
Avocado		3–5	
Grape	4–6		
Mandarin		4–6	
Nectarine	5–8		
Passionfruit		3–5	
Peach	2–6		
Pineapple, ripe		4–5	
Pineapple, green			4–5
Plum	2–7		
Semi-perishable (6–12 weeks)			
Coconut	8–12		
Orange		6–12	

Table 12 (continued)

Produce	Time at optimum temperature (weeks)		
	−1–4°C	5–9°C	10°C
Non-perishable (> 12 weeks)			
Apple	8–30		
Grapefruit			12–16
Lemon			12–20
Pear	8–30		
Vegetables			
Very perishable (0–4 weeks)			
Asparagus	2–4		
Bean	1–3		
Broccoli	1–2		
Brussels sprout	2–4		
Cauliflower	2–4		
Cucumber		2–4	
Lettuce	1–3		
Pea	1–3		
Rhubarb	2–3		
Spinach	1–2		
Sweet corn	1–2		
Tomato		4 days	1–3
Mushroom	2–3		
Perishable (4–8 weeks)			
Cabbage	4–8		
Semi-perishable (6–12 weeks)			
Celery	6–10		
Leek	8–12		
Marrow			6–10
Non-perishable (> 12 weeks)			
Beetroot	12–20		
Carrot	12–20		
Onion	12–28		
Parsnip	12–20		
Pumpkin			12–24
Potato		16–24	
Sweet potato			16–24
Swede turnip	16–24		

[1] Adapted from Hall. E.G. Mixed storage of foodstuffs. Sydney: CSIRO: 1973. Food Research Circular No. 9.
[2] Optimum storage temperature of green banana and tomato at all colour stages is 13°C.

must the required length of storage. In fruits and vegetables not liable to cold injury maximum storage life can be obtained by storage close to the freezing point of the tissue, but for chilling-sensitive produce the advantages of reduced respiration and fungal growth must be balanced against the potential losses from chilling injury. The storage life of produce is highly variable and can be related to the wide range of respiration rates among different tissues (Table 11). In general there is an inverse relation between respiration rate and storage life so that produce with a low respiration rate generally keeps longer. Tables of recommended temperatures and expected storage life have been compiled by various workers and agencies. Table 12 is an example. Published tables should be accepted only as a guide, since they are compiled from a large number of sources. The most suitable storage temperature for a particular commodity in any given locality should be determined from trials with the locally grown commodity, because climatic and soil factors may influence storage responses.

The examples given in Table 12 have been grouped according to the length of storage life that may be expected at optimum temperatures for each commodity. Generally those items which have the shortest storage life respire at high rates (most leafy vegetables), are harvested ripe (berry fruits), or are chilling sensitive (banana and cucumber).

A further major influence on storage life is susceptibility of produce to fungal decay. The growth of decay organisms is also slowed at low storage temperatures, but fresh produce gradually loses its natural resistance to the growth of decay organisms. Thus the duration of storage life is determined by the interaction between natural senescence (loss of quality), the growth of decay organisms and the susceptibility to low temperature injury.

COOLING OF PRODUCE

The object of cool storage is to restrict deterioration without causing abnormal ripening or other undesirable changes, thus maintaining the produce in a condition acceptable to the consumer for as long as possible. Of all foodstuffs that require low temperature storage, fruit and vegetables are the most demanding for both the engineer, who designs the cool store and the associated refrigeration equipment, and for the cool store operator. In addition to providing sufficient refrigeration capacity to cool the produce to the required temperature, provision must be made for continuous removal of the heat of respiration and maintenance of high relative humidities (Chapters 5 and 12). Cool stores for fresh produce are generally required to operate within relatively close temperature limits, both in space and time, to maximize storage life, to avoid freezing of produce and to minimize desiccation of produce.

The temperature of fruit and vegetables at harvest is close to that of ambient air and could be as high as 40°C. At this temperature respiration rate is extremely high, and storage life brief. It is often good practice to harvest early in the morning to take advantage of the lower temperatures generally prevailing at this time. Early morning harvesting may not be feasible for larger growers, and morning temperatures in tropical areas may still be relatively high. It is axiomatic that the quicker the temperature of the produce is reduced to the selected storage temperature the longer will be its storage life. Rapid cooling after harvest is generally referred to as precooling and particularly benefits the more perishable

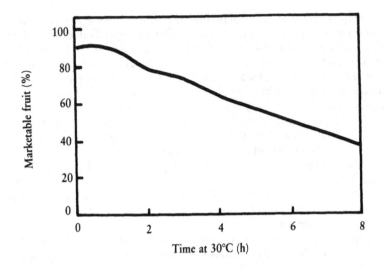

Figure 20   Effect of delay before cooling on quality of Shasta strawberries. (From Mitchell, F.G.; Guillou, R; and Parsons, R.A. With permission. Commercial cooling of fruits and vegetables. Berkeley: University of California Press; 1972. Manual number 43.)

or rapidly respiring fruits and vegetables. An example of the effect on quality of the delay in the cooling of strawberries is shown in Figure 20.

Rapid cooling of produce after harvest is often essential as refrigerated ships, land vehicles and containers are not designed to handle the full load of field heat, but are designed to merely maintain precooled produce at the selected carriage temperature. Rapid precooling was introduced in the early 1900s for long distance rail transport in the USA and for overseas fruit shipments. Maximum loading temperatures for refrigerated, perishable produce are now commonly supervised and closely controlled to ensure that produce will be discharged in good condition.

The term precooling is applied rather loosely in that it normally encompasses any cooling treatment given to produce before shipment, storage or processing. A stricter definition of precooling would include only those cooling methods by which the produce is cooled rapidly, and certainly within 24 hours of harvest. No legal definition of precooling has been established, so the definition must be sufficiently wide and flexible to embrace the cooling requirements of the various commodities in relation to their required postharvest life.

The method of cooling selected will depend greatly on the anticipated storage life of the commodity. Rapidly respiring commodities, which have a short post-harvest life, should be rapidly cooled immediately after harvest. Commodities that have a longer postharvest life generally do not have to be cooled quite so rapidly but should still be cooled as soon as possible. Commodities that are susceptible to chilling injury should be cooled according to their individual requirements, which may be as low as 10–12°C. The selection of the precooling method, therefore, depends on three main factors, the temperature of the produce at harvest, the physiology of the produce and the desired postharvest life. Produce that is to be 'cured' (Chapter 9) at temperatures above those required for extended storage is not normally precooled, for example, potato, yam and sweet potato.

Methods of cooling

Produce may be cooled by means of cold air (room cooling, forced-air cooling), cold water (hydrocooling), direct contact with ice, and evaporation of water from the produce (evaporative cooling, vacuum cooling). Fruit is normally cooled with cold air, although stone fruits benefit from hydrocooling. Any one of the cooling methods may be used for vegetables, depending upon the physiology and market requirements of individual vegetables.

Cooling rates

The rate of cooling of produce is dependent primarily upon:

1. rate of heat transfer from produce to cooling medium, which is especially influenced by rate of flow of the cooling medium around or into the containers of produce;
2. difference in temperature between the produce and the cooling medium;
3. the nature of the cooling medium; and
4. the thermal conductivity of the produce.

When hot produce is exposed to cool air kept at a constant temperature by refrigeration, the rate of cooling (°C per minute) is not constant, but diminishes exponentially as the temperature difference (driving force) between produce and air falls (Newton's Law). Because the rate of cooling varies, alternative ways of describing the cooling process are used. Two parameters are: (1) the cooling coefficient, defined as the ratio of the change in temperature per unit time at any moment to the difference in temperature between produce and air at the same moment; (2) the time required to reduce the temperature difference between produce and cooling medium by one half (Z) or by seven-eighths (S). Theoretically, Z and S are independent of the initial produce temperature and remain constant throughout the cooling period. S is more useful in commercial cooling operations because the temperature of the produce at seven-eighths cooling time is close to the required storage or transport temperature. In systems where the cooling rate is rapid the temperature change in the interior of produce lags considerably behind the change in surface temperature. This lag affects the relation between S and Z such that S may range from 2Z to 3Z. Mathematically seven-eighths cooling is expressed as:

$$S = \ln(8j)/C,$$

where j is the lag factor which may vary from 1 to 2 at the centre of cooling objects and C is the cooling coefficient, a negative value.

The cooling method, type of package, and the way the packages are stacked will all influence the rate of cooling of produce. The influence of these factors on Z value are shown in Tables 13 and 14.

Room cooling

Possibly the most common cooling technique is room cooling, where produce in boxes, cartons, bulk containers or other packages is exposed to cold air in a normal cool store. For adequate cooling, air velocities around the packages

Table 13   Half-cooling times for apples in 18 kilogram boxes[1]

Cooling method	Z(h)	
	Apples loose in box	Apples wrapped and packed
Conventional cool room	12	22
Tunnel, air velocity 200–400 m/min	4	14
High speed jet cooling, air velocity 740 m/min	0.75	
Hydrocooling (loose fruit)	0.33	
Single fruit: air velocity 40 m/min	1.25	
air velocity 400 m/min	0.5	

[1] Hall. E.G. Precooling and container shipping of citrus fruits. CSIRO Food Res. Q. 32: 1–10: 1972. With permission.

Table 14   Expected half-cooling times of centre fruits for various packages and stacking patterns[1]

Package	Z(h)		
	Freely exposed	Open stow[2]	Tight stow[2]
18 kg box loose fruit, unlidded	7	18	45
18 kg box wrapped and packed	23	35	45
Cell pack 18 kg carton	22	35	90
Tray pack 18 kg carton	23	43	90
Half-tonne bin, lidded, no floor vents	30	45	55
Half-tonne bin, lidded, vents 8% floor area	23	35	43
Half-tonne bin, open-top			
no floor vents	18		
5% floor vented	11		
10% floor vented	5		

[1] Hall, E.G. Precooling and container shipping of citrus fruits. CSIRO Food Res. Q. 32: 1–10: 1972. With permission.
[2] Units of 44 boxes or cartons stacked on pallets.

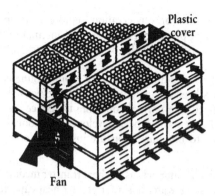

Figure 21 Air flow in forced-air (pressure) cooling. Placement of containers and proper use of baffles blocks air-return everywhere except through side vents in containers. Thus, air is forced to pass through containers and around produce to return to exhaust fans. As air is exhausted from the centre chamber, a slight pressure drop occurs across the produce. (From Mitchell, F.G.; Guillou, R.; and Parsons, R.A. (1972). With permission.)

should be at least 60 metres per minute. The design and operation of the system is simple (Chapter 12).

The produce may be cooled and stored in the same place thus requiring less rehandling, and peak loads on the refrigeration system are less than those of the faster cooling systems. Room cooling, however, suffers from the disadvantage that more space is required than is needed for good store management and the rate of cooling is relatively slow and thus may be inadequate for more sensitive produce. These disadvantages are especially pronounced when field bins or containers are unitized on pallets.

Forced-air (pressure) cooling

The rate of cooling with cold air may be significantly increased if the heat transfer surface is enlarged from that of the package to the total surfaces of the produce. By forcing the air through the packages and around each piece of produce, forced-air cooling can cool produce in about one-quarter to one-tenth the time required for room cooling; room cooling removes heat from only the surface of the package, the size and shape of the package being the limiting factor.

In the USA and Australia the most common method of forced-air cooling exposes produce in specially vented containers, to air at a higher pressure on one face of the container (Figure 21). The pressure differential between opposite faces ranges from barely measureable to about 250 pascals (25 millimetres water head) and air flows vary between 0.1 and 2.0 litres per second per kilogram. The speed of cooling can be adjusted by varying the rate of air flow. The refrigeration requirements for forced air cooling are often over-estimated because of a lack of understanding of the factors that limit the rate of heat loss from the cooling produce. Over-estimation of refrigeration requirements increases the initial capital cost of a cooling plant unnecessarily. The thermal properties, S and j, of stacks of packaged produce should be taken into account when calculating refrigeration capacity.

Maintaining high relative humidity during rapid cooling is not important but once the produce has been adequately cooled, air velocity should be reduced or the produce transferred to a normal coolroom to reduce the risk of desiccation. Various commercial systems have been developed that cool and humidify air used in forced-air systems to minimize moisture loss from cooled produce.

Hydrocooling

Hydrocooling, in which water acts as the heat transfer medium, is a rapid method for cooling produce, since water has a far greater heat capacity than air. Hydrocooling is rapid if the water contacts most of the surface of the produce and is maintained as close to 0°C as possible. In many hydrocooling systems, the produce passes under cold showers on a moving conveyor. Hydrocooling may also clean the produce, but contamination of the produce with spoilage micro-organisms occurs if soil is not removed from the system regularly and the water chlorinated. A further advantage is that the commodity loses little weight during hydrocooling.

Produce and containers must be tolerant to wetting. When cooling is completed, produce must be moved to a coolroom to prevent re-warming.

Contact icing

Before the advent of some of the more modern precooling techniques, contact or package icing was used extensively for precooling produce and maintaining temperature during transit, particularly for more perishable commodities such as

Figure 22   Precooled broccoli packed with crushed ice in an unvented polystyrene container.

leafy vegetables. Contact icing is now mainly employed as a supplement to other forms of precooling. The finely crushed ice or an ice slurry (liquid ice, 40 per cent water + 60 per cent ice + 0.1 per cent salt) is sprayed onto the top of the load inside the road or rail transit vehicle. This is often referred to as top icing.

In Australia, it is common practice to add crushed ice to cooled produce such as broccoli and sweetcorn packed in polystyrene containers to maintain freshness during transport and marketing (Figure 22).

Vacuum cooling

Vegetables that have a high surface to volume ratio may be rapidly and uniformly cooled by boiling off some of their water at low pressure. This technique is known as vacuum cooling (Figure 23) and is as rapid as hydrocooling. The produce is loaded into a sealed container, and the pressure is reduced to about 660 pascals (5 millimetres mercury). At this pressure water boils at 1°C, and the produce is cooled by evaporation of water from the tissue surface. For every 5°C drop in temperature, approximately 1 per cent of the produce weight is boiled off as water. This weight loss may be minimized by spraying the produce with water either before enclosing it in the vacuum chamber or towards the end of the vacuum cooling operation (hydrovacuum cooling).

The rate of cooling by this technique is largely dependent on the surface to volume ratio of the produce and the ease with which the produce loses water. Leafy vegetables are ideally suited to vacuum cooling; other vegetables, such as asparagus, broccoli, Brussels sprout, mushroom and celery, can also be successfully vacuum cooled. Fruits, which have a low surface to volume ratio and a waxy cuticle and therefore lose water slowly, do not benefit by vacuum cooling. Comparative cooling rates of several vegetables are shown in Table 15.

Table 15 Comparative cooling of vegetables under similar vacuum conditions[1]

Produce (initial temperature 20°C)	Final temperature (°C)
Lettuce, onion	2
Sweet corn	5
Broccoli	6
Asparagus, cabbage, celery, pea	7
Carrot	14
Potato, zucchini	18

[1] Adapted from American Society of Heating. Refrigerating and Air-Conditioning Engineers (1974).

Evaporative cooling

This is a simple process in which dry air is cooled by blowing it across a wet surface. Although the technique is restricted to regions with low relative humidity but with a good quality water supply, it has the advantage of low energy cost. The commodity may be cooled by either the humidified cool air or by misting with water and then blowing dry air over the wet fruit. The extent to which air may be cooled by evaporation of water is limited by the water holding capacity of the air (see Chapter 5).

Figure 23   Vacuum cooling of lettuce packed in fibreboard cartons. (Courtesy of Pennwalt Limited, Camberley, England.)

FURTHER READING

American Society of Heating, Refrigerating and Air-conditioning Engineers. ASHRAE Handbook of Refrigration Systems and applications. Atlanta, GA: 1986.

Burton, W.G. The physics and physiology of storage. Harris, P. M. ed. The potato crop. London: Chapman and Hall; 1978: 545–606.

Dossat, R.J. Principles of refrigeration. 2nd ed. SI version. New York: John Wiley and Sons; 1981.

Hallowell, E.R. Cold and freezer storage manual. 2nd ed. Westport, CT: AVI; 1980,

Hardenburg, R.E.; Watada, A.E.; Wang, C.Y. The commercial storage of fruits, vegetables and florist and nursery stocks, revised ed. Washington, DC: US Department of Agriculture; 1986. Agriculture Handbook No. 66.

International Institute of Refrigeration. Cooling and ripening of fruits in relation to quality. Paris; 1973. Commission C2.

Lipton, W.J.; Harvey, J.M. Compatibility of fruits and vegetables during transport in mixed loads. Washington, DC: US Department of Agriculture; 1977. Marketing Research Report No. 1070.

Mitchell, F.G.; Guillou, R.; Parsons, R.A. Commercial cooling of fruits and vegetables. Berkeley: University of California; 1972. Manual No. 43.

Porritt, S.W. Commercial storage of fruits and vegetables. Ottawa: Canada Department of Agriculture: 1974. Publication 1532.

Ryall, A.L.; Lipton, W.J. Handling, transportation, and storage of fruits and vegetables, revised ed. Vol. 1. Vegetables and melons. Westport, CT: AVI; 1979.

Ryall, A.L.; Pentzer, W.T. Handling, transportation and storage of fruits and vegetables, revised ed. Vol. 2. Fruits and tree nuts. Westport, CT: AVI; 1982.

Wade, N.L. Estimation of the refrigeration capacity required to cool horticultural produce. Int. J. Refrigeration. 7:358–66; 1984.

Whiteman, T.M. Freezing points of fruits, vegetables and florist stocks. Washington, DC: US Department of Agriculture; 1957. Marketing Research Report No. 196.

5
Effects of water loss and humidity

Fresh fruit and vegetables can be regarded as essentially water in fancy and expensive packages. This water was costly to put into those packages. Water loss is loss of saleable weight and thus is a direct loss in marketing. Measures that minimize water loss after harvest can be profitable. A loss in weight of only 5 per cent will cause many perishable commodities to appear wilted or shrivelled, and under warm, dry conditions this can happen to some produce in a few hours. Even in the absence of visible wilting, water loss can cause a loss of crispness, and undesirable changes in colour and palatability may ensue in some vegetables.

In contrast to conditions which promote water loss, conditions which result in wetting produce can result in disastrous losses with some commodities. Wetting can encourage the growth of rotting organisms and in some instances cause physical splitting of the commodity. Therefore, in studying the role of water in quality maintenance, it is necessary not only to understand the basic principles involved but also to examine the specific physiological responses of each commodity.

BASIC PRINCIPLES

Dry air is a mixture of about 78 per cent nitrogen and 21 per cent oxygen, with 0.034 per cent carbon dioxide, argon and other minor constituents comprising the remaining 1 per cent. Moist air consists of a mixture of dry air and water vapour. Humidity is the general term referring to the presence of water vapour in air. If water is placed in an enclosure containing dry air, water molecules will enter the vapour phase and the air will become saturated with water vapour. The amount of water vapour in air can vary from zero to a maximum that is dependent on temperature and pressure, for example, saturated air at 30°C comprises approximately 4 per cent water vapour. The evaporation of water is a physical process that requires energy.

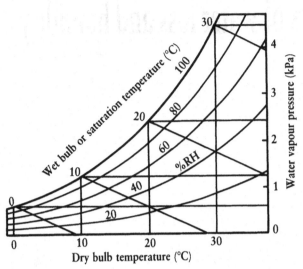

Figure 24 Simplified psychrometric chart.

Relative humidity (RH) is probably the best known term for expressing the humidity of moist air, and it is defined as the ratio of the water vapour pressure in the air to the saturation vapour pressure possible at the same temperature, expressed as a percentage. Saturated air, therefore, has a relative humidity of 100 per cent. When water-containing material, such as fruit, is placed in an enclosure filled with air, the water content of the air increases or decreases until equilibrium is reached. Equilibrium occurs when the number of water molecules entering and leaving the vapour phase is equal. The relative humidity at equilibrium is called the equilibrium relative humidity (ERH) which is a property of the material and its moisture content. Pure water has an ERH of 100 per cent.

The high water content of fruit and vegetables is held within the produce by osmotic forces within the cells, mostly as free water, although a small proportion is chemically bound and, therefore, is more tightly held and more stable. Water in plant tissues contains varying amounts of solutes which slightly depress the vapour pressure of the water. When fresh plant tissue is placed in an enclosure, the air will not become completely saturated due to the presence of these solutes and bound water. For most fresh produce an equilibrium relative humidity of at least 97 per cent is established.

The small reduction in equilibrium relative humidity due to the presence of solutes can also be related to the terms water activity and water potential. Water activity and water potential are expressions of the energy status of water within living tissues and this determines the availability of water for life processes. These measures are of particular interest to microbiologists, because small reductions in water activity and water potential equivalent to a reduction in equilibrium relative humidity to 95 per cent can inhibit the growth of many bacteria and fungi in culture, although growth of some pathogenic (disease-bearing) fungi is not prevented until water activity and water potential are reduced to less than the equivalent of 85 per cent ERH.

Psychrometric charts that graphically relate the various properties of moist air have been constructed. An abbreviated example is shown in Figure 24. The scale along the bottom axis indicates dry dulb temperatures as given by a wet-and-dry bulb hygrometer (Appendix 5). Dry air at all temperatures has zero water concentration and, therefore, zero water vapour pressure. The curved line at the top of the figure illustrates the relationship between vapour pressure and temperature in saturated air. By definition this is the line of 100 per cent relative humidity. Other curved lines can be drawn representing constant relative humidity over a range of temperatures. The curvature of these lines shows that the vapour pressure of water increases rapidly with temperature, for example, at 30°C in saturated air (100 per cent relative humidity) the vapour pressure of water is 4.3 kilopascals, at 20°C the vapour pressure is 2.4 kilopascals but at 10°C it is only 1.3 kilopascals.

The discrepancy in vapour pressure, or vapour pressure difference (VPD), which is the difference between the equilibrium relative humidity of the produce and the actual relative humidity of the air, has important consequences in the cooling of fresh produce. Even if saturated cool air is used to cool produce, as long as the produce remains warmer than the cooling air, it will lose water. Thus it is important to cool produce rapidly to minimize the vapour pressure difference between the produce and the cooling air stream.

Figure 24 also illustrates another important physical property of moist air, namely dewpoint. When moist air is cooled, a temperature is reached at which the vapour pressure of water reaches the maximum for that temperature. Water will then condense, for example, fog is formed or water condenses on a cooled surface. The temperature at which condensation occurs is the dewpoint temperature. The horizontal lines in Figure 23 represent dewpoint temperatures, thus air of 80 per cent relative humidity at 30°C becomes saturated when cooled to about 26°C. The dewpoint temperature is equal to the dry bulb temperature at the point of intersection with the saturation curve. The lines which slope upwards from right to left indicate constant wet bulb temperatures. Condensation has important consequences for cooled produce in packages when it is moved into warm moist air. Condensation can promote rots, weaken fibreboard packages and also accelerate warming of the produce. At low storage temperatures, where a high humidity is required, small fluctuations in temperature can result in excessive condensation on cooling surfaces and accentuated water loss from produce (Chapter 12).

FACTORS AFFECTING WATER LOSS

Surface area/volume effects

A major factor in the rate of water loss from produce is the surface area to volume ratio of the material. On purely physical grounds, there is a greater loss by evaporation from produce with a high surface area to unit volume ratio. Thus, other factors being equal, a leaf will lose moisture, and weight, much faster than a fruit, and a small fruit, or root, or tuber will lose weight faster than a large one.

Nature of surface coatings

The types of surfaces and underlying tissues of fruit and vegetables have a marked effect on the rate of water loss. Many types of produce have a waxy coating on the surface (cuticle) that is resistant to the passage of water or water vapour. Before harvest, the cuticle plays an essential part in maintaining within the tissues of the produce the high water content that is necessary for normal metabolism and growth by restricting water loss by evaporation.

The structure of the wax coating is more important than its thickness. Waxy coatings which consist of a complex, well ordered structure of overlapping plate-lets provide more resistance to the permeation of water than coatings which are thicker but flat and structureless. In the former, the water vapour must follow a more diverse path as it escapes into the atmosphere.

Underneath the waxy layer and the cuticle are the epidermal cells, which are compactly structured with minimal space between adjacent cells. The bulk of the movement of water vapour and other gases in and out of leaves is controlled by small pores called stomates (Figure 25), which are located at intervals in the epidermis. The stomates in harvested leafy produce normally close after a small amount of water is lost, but under some conditions, for example, rapid cooling of chilling-sensitive tissues, the stomates may remain open. Many fruit and storage organs have lenticels and not stomates. The outer layers in these products are often comprised of cork cells, which are toughened hypodermal cells packed

Figure 25 Open stomates on the surface of a passionfruit leaf. (Courtesy Dr J.M. Bain, formerly CSIRO Division of Food Research.)

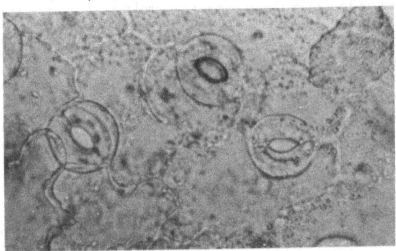

tightly together. Lenticels are narrow openings formed between the cork cells. There is no mechanism provided for the closing of lenticels. The rate of transpiration thus depends on the number and size of the pore openings and the nature of the wax coating. In mature fruit the lenticels are often blocked with wax and debris and are therefore non-functional. Thus loss of water as vapour, and respiratory gas exchange take place only by diffusion through the cuticle. The surface of some tubers and roots consists of periderm (cork), which comprises several layers of suberized cells. The periderm is generated from a layer of cambial cells.

Mechanical damage to tissue

Mechanical damage can greatly accelerate the rate of water loss from produce. Bruising damages the surface organization of the tissue and allows a much greater flow of gaseous material through the damaged area. Cuts are of even greater importance as they completely break the protective surface layer and directly expose the underlying tissues to the atmosphere. If the damage occurs early in the growth and development of the produce, the organ usually seals off the affected area with a layer of corky callus cells. The capacity for wound-healing generally diminishes as plant organs mature, so damage that occurs during harvesting or in postharvest operations remains unprotected. Some mature produce, notably tubers and roots, retain the capacity to seal off wound areas. Wound-healing in these products can be enhanced by curing at suitable temperatures and humidities (see Chapter 9). Damage to surface tissue can also occur following attack by pests or diseases and will result in increased rates of water loss.

CONTROL OF WATER LOSS

There is limited scope for modifying the tissue structure to reduce the rate of water loss. Therefore, methods of controlling the rate of water loss from the produce primarily involve lowering the capacity of the surrounding air to hold additional water by lowering the temperature or raising the humidity, that is, by reducing the vapour pressure difference between the produce and the air, or by providing a barrier to water loss. The effect of vapour pressure difference on water loss from apple cultivars is shown in Figure 26. The application of waxes and similar water-resistant skin coatings to the surface of produce can also reduce water loss (Chapter 10), as can appropriate packaging.

Increasing the humidity

An effective method for reducing water loss from fruit and vegetables is to increase the relative humidity of the air. This reduces the vapour pressure difference between the produce and the air and hence the amount of water required to be evaporated from the produce before the air is saturated with water vapour. The use of very high relative humidity, however, favours the growth of moulds and with some commodities, for example, citrus, can be conducive to rotting, however, the postharvest application of fungicides may overcome this

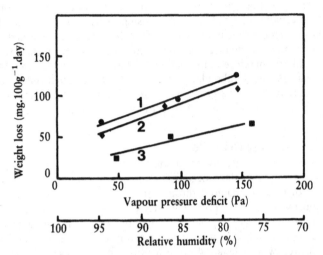

Figure 26   Effect of vapour pressure difference (deficit) on weight loss by the apple cultivars: 1. Golden Delicious; 2. Grimes Golden; and 3. Jonathan. (Adapted from Wells, A.W. Effects of storage temperature and humidity on loss of weight by fruit. Washington, D.C.; U.S. Department of Agriculture; 1962. Marketing Research Report No. 539.)

problem. It is a relatively simple procedure to increase the relative humidity of an air stream. This can be achieved by spraying water as a fine mist, by introducing steam or by increasing the temperature of the refrigeration coils. The addition of water vapour to a cold storage chamber can be controlled automatically with a humidistat.

For produce held in air of high relative humidity, the growth of microorganisms must be considered. In culture, most fungi cease to grow when the relative humidity is reduced to about 90 per cent and only a few can grow at 85 per cent. Under drier conditions spores cannot germinate and, even if there is enough free moisture in a wound to permit germination, a dry atmosphere may dry the exposed tissue fast enough to prevent infection and development of a rot. Experience has found that a relative humidity of 90 per cent is usually the best compromise condition for the storage of fruit, but a relative humidity of 98 to 100 per cent is better for leafy and some root vegetables which have a higher coefficient of transpiration. Onion and cucurbits, however, require a lower relative humidity, of 65 to 70 per cent, in storage to prevent excessive rotting. Storage of potatoes continuously at high humidity has a further advantage in that they are less prone to develop pressure bruises than if stored at low humidity.

Fruit ripen better, with not only better appearance because of the absence of shrivelling but also with better internal quality, at a relative humidity of at least 90 per cent. The necessity for controlling humidity in banana ripening rooms is generally well recognised. In contrast, storage of apple in very high humidities can be a disadvantage because reduced weight loss in cultivars susceptible to internal breakdown (low temperature breakdown) increases the incidence of this disorder (see Chapter 10).

Air movement

Air movement over the produce is a significant factor influencing the rate of moisture loss; air movement is required to remove heat from produce, but its effect on moisture loss must also be considered. There is always a microscopically thin layer of air (the boundary layer) adjacent to the surface of the produce in which the water vapour pressure is approximately in equilibrium with that of the produce itself. Air movement tends to sweep away this moist air from around the produce. Increasing the rate of air movement reduces the thickness of the boundary layer, increases the vapour pressure difference near the surface, and so increases the rate of moisture loss.

The faster that air moves across the surface of the fruit and vegetables the greater is the rate of water loss from the produce. In a cool store, restricting the air movement around the produce can, therefore, reduce the rate of water loss. After initial cooling, this can be achieved by decreasing the air movement generated by fans by running them at a lower speed or by reducing the length of time that they are operated. Open rooms with natural ventilation can be modified to restrict flow of air. Regulation of air movement, however, requires a compromise. Sufficient air movement is required to prevent large temperature gradients being produced within the storage chamber but at rates which will minimize water loss from the produce.

Packaging

Water loss can be reduced by placing a physical barrier around the produce to reduce air movement across its surface. The simplest methods are to cover stacks of produce with tarpaulins, or to pack the produce into bags, boxes, or cartons. Close packing of produce alone restricts the passage of air around individual items and thus water loss. Even placing the produce in mesh bages can have some beneficial effect, because there is a closer packing of individual items within the bags. More 'inner' fruits are created. These are protected from direct exposure to dry air by the outer layers in the bags.

The degree to which the rate of water loss is reduced is dependent on the permeability of the package to water vapour transfer as well as on the closeness of containment. All materials commonly used are permeable to water vapour to some extent. Materials such as polyethylene film can be considered to be relatively good vapour barriers, since their rate of water transfer is relatively low compared to that of paper and fibreboard, which have a high permeability to water vapour. Even the use of fibreboard packages or paperbags will substantially reduce water losses compared with unprotected, loose produce. But it must be remembered that packaging also reduces the rate of cooling by restricting air movement around individual items.

The use of very thin, heat-shrink plastic films for packaging individual fruits is a comparatively new technology that has significantly increased the storage life of fruit such as citrus, cucumbers and cantaloupes by greatly reducing the rate of water loss. Citrus and cantaloupes must be treated with approved fungicides before packaging in heat-shrink film because the saturated conditions under the film may encourage growth of pathogenic fungi. In the case of citrus fruit there is evidence that healing of skin injuries inflicted during harvesting and packing-house operations is promoted in heat-shrink film. The spread of decay from any fruit that become infected is prevented by film packaging. The film is not sealed and there is little change in the composition of the internal atmosphere in the fruit. Since the heat-shrink film is in direct contact with the surface of individual fruit there is little effect on the rate of heat exchange.

The ability of many packaging materials to absorb water must also be considered. Paper derivatives, jute (hessian) bags, and natural fibres generally, can absorb much water before becoming visibly damp. At the time of packing there is often a vapour pressure difference between the produce and the package, so that water is evaporated from the produce and absorbed by the packaging material. In the cool storage of apple and pear it has been found that a 'dry' wooden box weighing 4 kilograms can absorb about 500 grams of water at 0°C. Packages should be equilibrated at high humidity before use, but this is considered impractical commercially. An alternative procedure is to waterproof the packaging material by the incorporation of waxes or resins. Such packages are available commercially, but are necessarily more expensive than the untreated materials.

FURTHER READING

American Society for Horticultural Science. Relative humidity—physical realities and horticultural implications: proceedings of the symposium; 14 October 1977; Salt Lake City, UT, HortScience 13: 549–74: 1978.

Beek, G. van. Practical applications of transpiration coefficients of horticultural produce. ASHRAE Trans. 91; Part 1B: 708–25; 1985.

Ben-Yehoshua, S. Transpiration, water stress, and gas exchange. Weichmann, J. ed. Postharvest physiology of vegetables. New York: Marcel Dekker; 1987: 113–70.

Duckworth, R.B. Fruit and vegetables. Oxford: Pergamon Press: 1966.

Fockens, F.H.; Meffert, H.F. Th. Biophysical properties of horticultural products as related to loss of moisture during cooling down. J.Sci. Food Agric. 23: 285–98; 1972.

Gaffney, J.J.; Baird, C.D.; Chau, K.V. Influence of air flow rate, respiration, evaporative cooling, and other factors affecting weight loss calculations for fruit and vegetables. ASHRAE Trans. 91, Part 1B: 690–707: 1985.

Grierson, W.; Wardowski, W.F. Humidity in horticulture. HortScience 10: 356–60; 1975.

Grierson, W.; Wardowski, W.F. Relative humidity effects on the postharvest life of fruits and vegetables. HortScience 13: 570–4; 1978.

Hruschka, H.W. Postharvest weight loss and shrivel in five fruits and five vegetables. Washington, DC: US Department of Agriculture; 1977. Marketing Research Report No. 1059.

Sastry, S.K. Factors affecting shrinkage of foods in refrigerated storage. ASHRAE Trans. 91, Part 1B: 683–9; 1985.

Sastry, S.K. Moisture losses from perishable commodities: recent research and development. Int. J. Refrig. 8: 343–6; 1985.

Wells, A.W. Effects of storage temperature and humidity on loss of weight by fruit. Washington, DC: US Department of Agriculture: 1962. Marketing Research Report No. 539.

6
Storage atmosphere

The composition of gases in the storage atmosphere can affect the storage life of produce. Alteration in the concentrations of the respiratory gases, oxygen and carbon dioxide, may extend storage life. This is generally used as an adjunct to low temperature storage, but for some commodities modification of the storage atmosphere can usefully substitute for refrigeration. Many volatile compounds, evolved by produce and from other sources, may accumulate in the storage atmosphere. Ethylene is the most important of these compounds. Its accumulation above certain critical levels may reduce storage life, so methods for its removal become important. Carbon monoxide (CO) is not evolved by fresh produce, but it may be introduced to storage atmospheres by equipment powered by internal combustion engines. Carbon monoxide may reach levels toxic to persons working in the storage chambers and with some produce may give effects which mimic those induced by ethylene. But there are examples of beneficial responses to added carbon monoxide, namely the control of butt discolouration and retardation of the growth of Botrytis rots in lettuce.

The terms controlled atmosphere (CA) storage, modified atmosphere (MA) storage and 'gas' storage are frequently used. These terms imply the addition or removal of gases resulting in an atmospheric composition different from that of normal air. Thus the levels of carbon dioxide, oxygen, nitrogen, ethylene and carbon monoxide in the atmosphere may be manipulated. Controlled atmosphere storage generally refers to decreased oxygen and increased carbon dioxide concentrations and implies precise control of these gases, whereas the term modified atmosphere storage is used when the composition of the storage atmosphere is not closely controlled, for example, in plastic film packages, and where the change in the composition of the atmosphere arises intentionally or unintentionally. The original term, gas storage, is now considered inappropriate and should not be used.

CARBON DIOXIDE AND OXYGEN

The general equation for produce respiration is:

$$glucose + oxygen \rightarrow carbon\ dioxide + water$$

It suggests that respiration could be slowed by limiting the oxygen or by raising the carbon dioxide concentration in the storage atmosphere. The principle appears to have been applied in ancient times, even if unwittingly. The earliest use of modified atmosphere storage may possibly be attributed to the Chinese. Ancient writings report that litchis were transported from southern China to northern China in sealed clay pots to which fresh leaves and grass were added. It may be surmised that during the two-week journey, respiration of the fruit, leaves, and grass generated a high carbon dioxide-low oxygen atmosphere in the pots which retarded ripening of the litchis. Other examples of primitive modified atmosphere storage include the burying of apples in the ground and the carriage of fruit in the unventilated holds of ships. Dalrymple reports that the first scientific observations of the effects of atmospheres on fruit ripening were made in 1819–20 by Jacques Berard, a Professor of chemistry at Montpellier in France. Several further independent studies of the effects of controlled atmospheres on fruit ripening were made in the USA. One study involved the construction of a primitive controlled atmosphere store by Nyce in Cleveland, USA, in which apples were successfully stored. However, it was not until the work of Kidd and West at the Low Temperature Research Station at Cambridge, UK, that a sound basis for the controlled atmosphere storage of produce was established.

Kidd and West extended the earlier findings, that the composition of the atmosphere affected the metabolic rate of plant tissues, by showing that high carbon dioxide and low oxygen concentrations lowered the respiration rate of seeds and delayed their germination. It was soon apparent to them that there should be a similar effect in fruit. They subsequently concentrated their work on apple, which culminated in the publication in 1927 of the classic bulletin entitled *Gas Storage of Fruit*.

An important stimulus to the work of Kidd and West was the fact that the commercial cultivars grown in the UK were subject to low temperature disorders when stored at less than 3°C and that storage life was too short at temperatures above 3°C. During the last fifty years the effects of controlled atmosphere and modified atmosphere have been extensively tested on a wide range of produce, but the responses have varied considerably. Despite extensive research, major commercial application of controlled atmosphere has been confined to some apple and pear cultivars, but modified atmosphere has been applied successfully during transport of some produce. High carbon dioxide levels have been used primarily as a fungistat during the transport of strawberries. Improved out-turns of lettuce have been achieved by flushing rail trucks or containers with nitrogen and addition of up to 8 per cent carbon monoxide. Potential commercial applications of naturally generated modified atmospheres exist for some other fruits, for example banana, avocado and mango.

Factors which have influenced the adoption of controlled or modified atmospheres for different commodities include:

1. Inherent storage life in air. If the produce can be stored in a satisfactory condition in air for the total marketing period desired, then there is no need to resort to other storage techniques, such as controlled atmosphere, to prolong storage life.
2. Existence and magnitude of a favourable response to controlled atmosphere or modified atmosphere. There must be a distinct beneficial effect. Not all produce responds favourably to atmosphere regulation and some produce is little affected.
3. Seasonal availability. Use of atmospheres can be advantageous where produce is harvested over a relatively short period in the year. Maximum storage life of such produce is often desirable to extend the marketing period as far as possible over the whole year.
4. Value of the commodity in relation to the additional cost of controlled atmosphere or modified atmosphere. There needs to be a distinct financial gain from the use of atmosphere control.
5. Availability of substitute commodities. While produce may be stored satisfactorily in controlled atmosphere, it may be more economical to import produce from another region or country that has a different harvest period.

Metabolic effects

Increases in carbon dioxide and decreases in oxygen concentrations exert largely independent effects on respiration and other metabolic reactions. Generally, the oxyen concentration must be reduced to less than 10 per cent before any retardation of respiration is achieved. For apples stored at 5°C, the oxygen level must be reduced to about 2.5 per cent to achieve a 50 per cent reduction in respiration rate. Care must also be taken to ensure that sufficient oxygen is retained in the atmosphere so that anaerobic respiration, with its associated development of off-flavours, is not initiated.

The reduction in the concentration of oxygen necessary to achieve a retardation of respiration is dependent on the storage temperature. As the temperature is lowered the required concentration of oxygen is also reduced. The critical level of oxygen at which anaerobic respiration occurs is determined mainly by the rate of respiration and is, therefore, greater at higher temperatures. Tolerance to low oxygen levels varies considerably among different commodities. The critical level of oxygen may vary with the time of exposure, with lower levels being tolerated for shorter periods. It may also be affected by the level of carbon dioxide, since lower levels of oxygen often seem to be better tolerated when carbon dioxide is absent or at a low level.

The addition of only a few per cent of carbon dioxide to the storage atmosphere can have a marked effect on respiration. But if carbon dioxide levels are too high, effects similar to those caused by anaerobiosis (lack of oxygen) can be initiated. Responses to increased carbon dioxide levels vary even more widely than responses to reduced oxygen: cherries and strawberries will withstand, and even benefit from, exposure to 30 per cent carbon dioxide for short periods; some

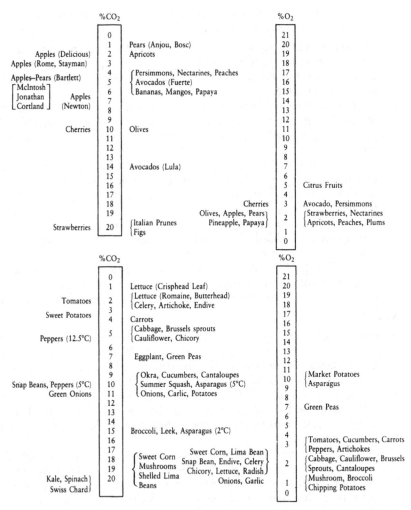

Figure 27 Relative tolerance of fruits and vegetables to elevated carbon dioxide and reduced oxygen levels at recommended storage temperatures. Normal atmospheric air comprises 0.034 per cent carbon dioxide, 21 per cent oxygen and about 79 per cent nitrogen. (From Kader, A.A.; Morris, L.L. Relative tolerance of fruits and vegetables to elevated CO_2 and reduced O_2 levels. Dewey, D.H., ed. Controlled atmospheres for the storage and transport of horticultural crops. Michigan State University, East Lansing, MI; 1977: 260–5.)

apple cultivars are injured by 2 per cent oxygen in storage; and many vegetables appear to respond best to low oxygen when carbon dioxide is kept low or is absent (Figure 27).

Many of the beneficial results of controlled atmosphere and modified atmosphere cannot simply be attributed to a reduction in respiration. For example, under ideal experimental conditions a twelve-fold increase in the storage life of green banana can be achieved by ventilating the fruits with an atmosphere comprising 5 per cent carbon dioxide, 3 per cent oxygen and 92 per cent nitrogen in the absence of ethylene, but respiration measured in terms of oxygen uptake is reduced to only one quarter of the rate in air. The greatly increased storage life is attributed to a reduction in the rate of natural ethylene production by the bananas and also to a reduced sensitivity of the fruits to ethylene.

In green vegetables, improved retention of green colour in low oxygen atmospheres is due mainly to a lowering of the rate of chlorophyll destruction. An interesting and contrasting effect has been noted in potato. Greening due to exposure to light can be prevented for several days by maintaining tubers in an atmosphere containing about 15 per cent carbon dioxide.

Controlled atmospheres, particularly those containing high carbon dioxide, inhibit breakdown of pectic substances so that a firmer texture is retained for a longer period. Retention of flavour may also be improved. But the responses of various commodities to controlled atmospheres can be conflicting. For example, increased carbon dioxide aids in retention of organic acids in tomato but accelerates loss of acids in asparagus. Maturity at harvest is more critical for controlled atmosphere storage than for ordinary air storage. Because of the widely varying responses of different commodities, and among cultivars, to alterations in oxygen and carbon dioxide concentrations, ideal combinations need to be determined experimentally for each commodity.

Effect on microbial growth

The activity of several decay organisms can be reduced by atmospheres containing 10 per cent carbon dioxide or more, provided that the commodity is not injured by such high carbon dioxide levels. Since strawberries can tolerate high carbon dioxide, transport of strawberries under modified atmosphere has been found to significantly reduce rotting and give a valuable extension in marketing life and quality (Table 16).

Table 16 Decay of strawberries as influenced by carbon dioxide concentration[1]

Storage condition	air 0% CO_2	10% CO_2	20% CO_2	30% CO_2
	Decayed strawberries (%)			
3 Days at 5°C in storage atmosphere	11.4	4.5	1.7	1.3
Plus 1 day at 15°C in air	35.4	8.5	4.7	4.0
Plus 2 days at 15°C in air	64.4	26.2	10.8	8.3

[1] Adapted from Harris. C.M.: Harvey. J.M. Quality and decay of California strawberries stored in carbon dioxide-enriched atmospheres. Plant Dis. Reptr 57: 44–6: 1973.

Many commodities cannot tolerate high carbon dioxide levels so that, in practice, atmosphere control cannot always be relied on to reduce rotting. Nevertheless, atmosphere control (by carbon dioxide and oxygen) may reduce rotting of produce by retarding ripening and senescence, since the resistance of the produce host to pathogens decreases as it ripens or ages. In contrast, some fruits, such as banana and mango, respond well to atmosphere control but eventually lose their resistance to the latent anthracnose disease which is then the factor limiting storage life. Controlled atmosphere or modified atmosphere storage does not necessarily retard ageing and the loss of resistance to decay organisms at the same rates.

ETHYLENE

The commencement of natural ripening in climacteric fruits is accompanied by an increase in ethylene production. Treatment of pre-climacteric fruits with exogenous ethylene advances the onset of ripening. This response is used widely in commercial practice to achieve controlled ripening of fruits such as the banana (Chapter 10).

Ethylene produced by other fruits or from outside sources may accumulate in a storage chamber containing unripe fruits to a level that initiates unwanted ripening. Often during marketing, several types of fruit and vegetables are stored together, and under these conditions ethylene given off by one commodity can adversely affect another. Coal gas, petroleum gas and exhaust gases from internal combustion engines contain ethylene, and contamination of stored produce by these gases may introduce sufficient ethylene to initiate ripening in fruit, promote senescence in leafy tissues and, if present in high enough concentrations, reduce the storage life of many commodities.

In addition to methods which reduce the ethylene concentration in the storage atmosphere, the sensitivity of produce to ethylene may also be lessened by storage at low temperature, and by either raising the level of carbon dioxide or decreasing the level of oxygen. Under these conditions the amount of ethylene required to induce ripening is increased. In practice, ripening or senescence is delayed by maintaining ethylene at low levels in storage rooms. This can be achieved by storing ripe and unripe produce in separate storage rooms and by ensuring that gas pipes, gas cylinders and exhaust gases from internal combustion engines are kept well away from storage rooms.

Other physical and chemical methods are available for reducing ethylene levels in storage rooms. The simplest method may be to ensure good ventilation of the storage chamber with air from outside the storage complex. The ethylene concentration in the atmosphere is normally less than 0.01 microlitre per litre, unless there is contamination from nearby industrial sources or heavy automobile traffic. Ventilation with external air could be applicable where no large temperature differential exists between the external air and the air in storage chamber. If there is a large temperature difference it may be necessary to cool the air before admitting it to the chamber.

Ethylene in the atmosphere can be destroyed by oxidation to carbon dioxide and water. Ozone (O_3) is a suitable oxidizing agent for destroying ethylene as it is generated readily from atmospheric oxygen by an electric discharge or by ultraviolet radiation, and since it is gaseous, it readily mixes with ethylene. Some precautions must be taken with ozone: it is a reactive substance and will corrode metal pipes and fittings in refrigeration equipment and react with paper products used to package the produce; it readily injures produce and can be toxic to man at relatively low concentrations. The widespread use of ozone has been hampered by difficulties in controlling its concentration. These difficulties with ozone have been overcome by utilizing ultraviolet radiation at two specific wavelengths. By manipulating the relative intensities of radiation at 184 nanometres and 254 nanometres admitted to a reaction chamber (Figure 28), atomic oxygen (O) is produced instead of ozone. Atomic oxygen is more reactive than ozone and reacts more rapidly with ethylene and other volatile compounds introduced into the reaction chamber. Excess atomic oxygen is rapidly converted to oxygen (O_2). More recently it has been found that excess ozone produced by radiation at 184 nanometres can be removed by passing the storage atmosphere as it leaves the reaction chamber through a second chamber containing rusting steel wool. This reduces the need for radiation at 254 nanometres and saves energy.

Ethylene can also be destroyed by potassium permanganate ($KMnO_4$), a strong oxidizing agent. Since potassium permanganate is non-volatile, it can be separated from produce, thus eliminating the risk of chemical injury. To ensure efficient destruction of ethylene, it is necessary to expose the storage atmosphere to a large surface area of potassium permanganate. This can be achieved by coating an inert inorganic porous support, such as a mixture of cement and expanded mica, with a saturated solution of potassium permanganate. Potassium permanganate applied in this manner has been found to retard the ripening of banana and avocado when used in conjunction with modified atmosphere storage in polyethylene bags (Table 17). The high carbon dioxide and low oxygen atmosphere generated within the sealed bags decreases the response by the fruits to ethylene and hence retards ripening. The addition of potassium permanganate further retards ripening by maintaining ethylene at a low level for a long period; thus

Figure 28 System for removal of ethylene with atomic oxygen generated by ultra violet radiation. (Derived from Scott, K.J.; Wills, R.B.H. Atmospheric pollutants destroyed in an ultra violet scrubber. Lab. Practice 22: 103–6; 1973.)

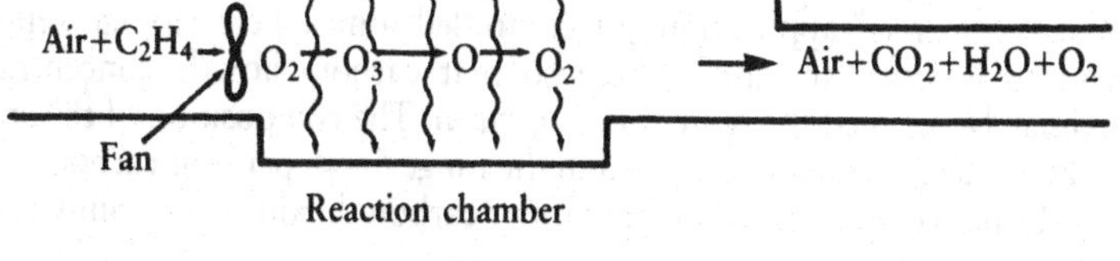

Reaction chamber

Table 17 Shelf life of banana and avocado held at 20°C in modified atmosphere with potassium permanganate

Treatment	Shelf life (days)
Air control	up to 7
Sealed polyethylene bags	14
Sealed bags—potassium permanganate	21

storage or transport life is prolonged. Fruits must be removed from the bags to achieve normal ripening. If held for longer periods under modified atmosphere, the fruits may not ripen satisfactorily after removal; the development of anthracnose infection, if not controlled, may also shorten the storage life.

Sealed polyethylene bags plus potassium permanganate will also delay ripening of whole bunches of bananas after harvest (Figure 29). The technique has been used successfully to delay the ripening of whole bunches of bananas during growth on the plant. The technique of modified atmosphere and ethylene absorption may be applicable to other fruits and vegetables which tolerate wide variations in carbon dioxide and oxygen levels, for example, avocado and mango. The advantages of the technique are its simplicity and low cost, although ripe rot or anthracnose infection can be a problem. The development of anthracnose on banana can be controlled with benzimidazole fungicides.

Ethylene at concentrations of more than 0.1 microlitre per litre accelerates senescence of non-climacteric fruits, vegetables and cut flowers and may induce physiological disorders and undesirable flavours in some vegetables. The storage of lemons provides a practical example of the benefits of maintaining low ethylene in the storage atmosphere. Under experimental conditions, when ethylene was removed from the storage atmosphere, lemons stored in air had a better appearance, higher juice yield, and higher acid levels than if ethylene were present in the storage atmosphere. When ethylene removal was combined with controlled atmosphere storage, mould wastage was reduced and retention of green colour in the peel enhanced. Although citrus fruits produce little ethylene under normal conditions, this small amount plus that evolved by moulds, eg *Penicillium digitatum* (green mould), can accumulate to physiologically active levels in the confined atmosphere of storage chambers.

METHODS FOR MODIFYING CARBON DIOXIDE AND OXYGEN CONCENTRATIONS

The first commercial applications of controlled atmosphere storage relied on the fruit to generate the atmosphere so that carbon dioxide concentrations approximately equalled the reduction in oxygen. The composition of the storage atmosphere was generally maintained in the range 5–10 per cent carbon dioxide and 16–11 per cent oxygen. The buildup of carbon dioxide was mainly respon-

Figure 29   Retardation of the ripening of a whole bunch of bananas by sealing the bunch in a polyethylene bag with potassium permangante. The bunches were held for thirty-seven days at 20°C, in which time the untreated bunch (left) has ripened and rotted; in contrast the treated bunch (right) is still hard green. (Courtesy K.J. Scott, NSW Department of Agriculture.)

sible for the increased storage life. The store was ventilated regularly to maintain the required carbon dioxide level. Further research showed that low oxygen levels were of greater benefit and that some important cultivars of apple and pear were sensitive to carbon dioxide levels above 3 per cent. To maintain low oxygen levels in the store atmosphere, it was necessary to make existing cool stores more gas-tight and to recirculate part of the store atmosphere through a scrubber to remove excess carbon dioxide. Improved cool stores for controlled atmosphere storage have now been developed. These are more gas-tight and less prone to moisture condensation in the insulation at the high relative humidities required for produce storage (Chapter 12). Recent development of external generators (Chapter 12) has further increased the attractiveness of controlled atmosphere storage; these generators burn fuels such as propane or petroleum gas and rapidly reduce the oxygen in the store to a required low level which is then maintained. Excess carbon dioxide is removed by scrubbing equipment. Gas generators also allow controlled atmosphere storage to be accomplished, albeit expensively, in a store that is not completely gas-tight.

As a result of these developments in cool store design and operation, the more effective low oxygen atmospheres containing 2–5 per cent carbon dioxide and 2–3 per cent oxygen can readily be maintained. The oxygen concentration must be carefully controlled to prevent anaerobic respiration or fermentation in this type of atmosphere. Methods of constructing and maintaining controlled atmosphere stores are discussed in Chapter 12.

Atmosphere control by the addition of nitrogen and carbon dioxide

In recent years there has been renewed interest in atmosphere control in the long distance transport of perishable produce in containers. One of the factors responsible for this interest is the availability in some countries, notably the USA, of cheap liquid nitrogen. Atmosphere control in large containers has involved either the continuous introduction of carbon dioxide or nitrogen gas during the journey, or charging the container with the appropriate atmosphere before the journey, with no further introduction of gas. Carbon dioxide or nitrogen from cylinders are commonly used, depending on whether the requirement is for high carbon dioxide or low oxygen, or both. The advent of hollow fibre systems for separating oxygen and nitrogen in air has provided a new means of producing low oxygen atmospheres for continuous ventilation of produce in large containers.

The use of liquid nitrogen as a refrigerant in the transport of perishables had stimulated interest in the responses of fruit and vegetables to 100 per cent nitrogen and to low levels of oxygen, both under refrigeration and at higher temperatures. As early as 1963, Ryall studied the effects of low oxygen levels on several commodities. Lettuce and strawberries withstood several days in 100 per cent nitrogen at 0°C, but spinach had a bitter flavour after four days. Green bananas and tomatoes in 100 per cent nitrogen at 15°C for four to seven days did not ripen and failed afterwards to ripen at 21°C in air. Both withstood 1 per cent oxygen: 99 per cent nitrogen as did peaches, but subsequent ripening was retarded. Ryall concluded that liquid nitrogen refrigeration of road and rail transport vehicles was unlikely to be damaging to most perishables, because the likelihood of maintaining oxygen-free atmospheres was remote and even 1 per cent oxygen was enough for most commodities to remain viable for several days.

Storage in plastic films

The use of plastic films to achieve modified atmosphere is increasing rapidly, not only in packaging but also in controlled atmosphere store construction (Chapter 12). Polyethylene box liners, either sealed or unsealed, have been used for several years in the storage of pear and apple and to a lesser extent of other produce. Unsealed or perforated bags are commonly used to minimize weight loss and reduce abrasion damage. A major problem with sealed bags is that the atmosphere inside depends on the temperature, because the permeability of the film to gases is virtually independent of temperatures at which produce is normally handled, whereas respiration is temperature-dependent. Thus sealed bags are risky when the temperature varies more than a few degrees, unless the produce has a low rate of respiration, or is tolerant to atmospheres which vary widely in carbon dioxide and oxygen concentration (like the banana), or both. The film used is generally 0.04 millimetre (0.0015 inch) low density polyethylene. To avoid brown spot and other carbon dioxide injuries, sachets of fresh hydrated lime can be included in the bag to reduce the carbon dioxide concentration. At the rate of 100–200 grams per 10 kilograms of fruit, this has proved useful with apple and pear, particularly with carbon dioxide-sensitive cultivars in cool storage.

The attainment of modified atmosphere in polyethylene bags filled with produce can be accelerated by evacuating the bags to between 50 and 85 kilopascals (380–635 millimetres mercury) and then sealing them. Since the polyethylene film is permeable to nitrogen, oxygen and carbon dioxide, the pressure inside returns to atmospheric pressure, but the initial rapid reduction of oxygen concentration is often useful. Eventually the composition of the atmosphere approaches that in bags not subjected to initial evacuation.

The availability of a range of new food-grade polymeric films with different permeabilities to the atmospheric gases has revived interest in packaging in sealed bags. The use of such films for refrigerated produce offers a cheaper alternative to using large containers equipped to provide MA or CA.

Hypobaric storage

Hypobaric storage is a form of controlled atmosphere storage in which the produce is stored in a partial vacuum. The vacuum chamber is vented continuously with water saturated air to maintain oxygen levels and to minimize water loss. Ripening of fruit is retarded by hypobaric storage, due to the reduction in the partial pressure of oxygen and for some fruits also to the reduction in ethylene levels. A reduction in pressure of air to 10 kilopascals (0.1 atmosphere) is equivalent to reducing the oxygen concentration to about 2 per cent at normal atmospheric pressure. Hypobaric stores are expensive to construct because of the low internal pressures required, and this high cost of application appears to limit hypobaric storage to high value produce such as cut flowers.

FURTHER READING

Bixler, H.J.; Sweeting, O.J. Barrier properties of polymer films. Sweeting, O.J. ed. The science and technology of polymeric films. Vol 2. New York: Wiley—Interscience; 1971: 1–130.

Dalrymple, D.G. The development of an agricultural technology: controlled-atmosphere storage of fruit. Technol. Culture 10: 35–48; 1969.

Dewey, D.H.; Herner, R.C.; Dilley, D.R., eds. Controlled atmospheres for storage and transport of horticultural crops. Proceedings of the national controlled atmosphere research conference: Michigan State University, East Lansing, MI; 27–28 January 1969. East Lansing, MI: 1969. Michigan State University Report No. 9.

Hardenburg, R.E. Effect of in-package environment on keeping quality of fruits and vegetables. HortScience 6: 198–201; 1971.

Irving, A.R. Transport of horticultural produce under modified atmospheres. CSIRO Food Research Q. 44: 25–33; 1984.

Isenberg, F.M.R., ed. Symposium on vegetable storage. Acta Hortic. 62: 1–361; 1976.

Isenberg, F.M.R. Controlled atmosphere storage of vegetables. Horticultural Reviews Vol. 1. Westport, CT: AVI; 1978: 337–94.

Kader, A.A. Modified atmospheres. An indexed reference list with emphasis on horticultural commodities. Davis, CA: University of California: 1985. Supplement no. 4, Postharvest Horticulture Series 3.

Kader, A.A. Biochemical and physiological basis for effects of controlled and modified atmospheres on fruits and vegetables. Food Technol. 40 (5): 99–104; 1986.

Kader, A.A.; Morris, L.L. Modified atmospheres. An indexed reference list with emphasis on horticultural commodities. Davis, CA: University of California; 1977. Supplement no. 2, Vegetable Crops Series 187.

Kader, A.A.; Morris, L.L. Modified atmospheres. An indexed reference list with emphasis on horticultural commodities. Davis, CA: University of California; 1981. Supplement no. 3, Vegetable Crops Series 213.

Kidd, F.; West, C; Kidd, M.N. Gas storage of fruit. London: Great Britain Department of Scientific and Industrial Research; 1927. Food Investigation Board Special Report No. 30.

MacLean, D.L.; Stookey, D.J.; Metzger, T.R. Fundamentals of gas permeation. Hydrocarbon processing. Int. Ed. 62: 47–51; 1983.

Mermelstein, N.H. Hypobaric transport and storage of fresh meats and produce earns 1979 IFT Food Technology Industrial Achievement Award. Food Technol. 33(7): 32–40: 1979.

Morris, L.L.; Claypool, L.L.; Murr, D.P. Modified atmospheres. An indexed reference list through 1969, with emphasis on horticultural commodities. Berkeley, CA: University of California; 1971.

Murr, D.P.; Kader, A.A.; Morris, L.L. Modified atmospheres. An indexed reference list with emphasis on horticultural commodities. Davis, CA: University of California: 1974, Supplement no. 1, Vegetable Crop Series 168.

Paine, F.A.; Paine, H.Y.A handbook of food packaging. p305. Glasgow: Leonard Hill; 1983.

Ryall, A.L. Effects of modified atmospheres from liquified gases on fresh produce. Proceedings of seventeenth national conference on handling of perishable agricultural commodities: Purdue University, Lafayette, IN: 11–14 March 1963. Washington, DC: US Department of Agriculture; 1963.

Scott, K.J.; McGlasson, W.B.; Roberts, E.A. Potassium permanganate as an ethylene absorbent in polyethylene bags to delay ripening of bananas during storage. Aust. J. Exp. Agric. Anim. Husb. 10: 237–40: 1970.

Sherman, M. Control of ethylene in the postharvest environment. HortScience 20: 57–60; 1985.

Shorter, A.J.; Scott, K.J. Removal of ethylene from air and low oxygen atmospheres with ultra violet radiation. Lebensm.—Wiss.u.—Technol. 19: 176–9; 1986.

Smock, R.M. Controlled atmosphere storage of fruits. Horticultural Reviews Vol. 1, Westport, CT: AVI; 1978: 301–36.

Stoll, K. The storage of vegetables in controlled atmospheres. Bull. Inst. Int. Froid 54: 1302–24; 1974.

Tomkins, R.G. The conditions produced in film packages by fresh fruits and vegetables and the effect of these conditions on storage life. J. Appl. Bacteriol. 25: 290–307; 1962.

7
Physiological disorders

Physiological disorders refer to the breakdown of tissue that is not caused by either invasion by pathogens (disease-causing organisms) or by mechanical damage. They may develop in response to an adverse environment, especially temperature, or to a nutritional deficiency during growth and development.

LOW TEMPERATURE DISORDERS

Storage of produce at low temperature is beneficial, because the rate of respiration and metabolism is reduced (Chapter 3). Low storage temperatures do not, however, suppress all aspects of metabolism to the same extent. Some reactions are sensitive to low temperature and cease completely below a critical temperature. Several such cold-labile enzyme systems have been isolated from plant tissue. Decreasing temperature does not reduce the activity of other systems to the same extent as it does respiration. For these systems, this leads to an accumulation of reaction products and possibly a shortage of reactants, while the converse occurs with cold-labile systems. The overall effect is that an imbalance in metabolism is created, and if it becomes serious enough to result in an essential substrate not being provided, or toxic products being accumulated, the cells will cease to function properly and will eventually lose their integrity and structure. These collapsed cells manifest themselves as areas of brown tissue in the produce. Metabolic disturbances occurring at reduced temperature are generally divided into two main groups: chilling injury and physiological disorders.

Chilling injury

Chilling injury is a disorder which has long been observed in plant tissues, especially those of tropical or subtropical origin. It results from the exposure of susceptible tissues to temperatures below 15°C, although the critical temperature at which chilling injury symptoms are produced varies among different commodities.

Table 18　Chilling injury symptoms of some fruits

Produce	Approx. lowest safe storage temperature (°C)	Symptoms
Avocado	5–12[1]	Pitting, browning of pulp and vascular strands
Banana	12	Brown streaking on skin
Cucumber	7	Dark coloured, water-soaked areas
Egg-plant	7	Surface scald
Lemon	10	Pitting of flavedo, membrane staining, red blotches
Lime	7	Pitting
Mango	5–12[1]	Dull skin, brown areas
Melon	7–10[1]	Pitting, surface rots
Papaya	7	Pitting, water-soaked areas
Pineapple	6–12[1]	Brown or black flesh
Tomato	10–12[1]	Pitting, Alternaria rots

[1] A range of temperature indicates variability between cultivars in their susceptibility to chilling injury.

Chilling injury is a separate phenomenon from freezing injury, which results from the freezing of the tissue and formation of ice crystals at temperatures below the freezing point. A clear distinction can, therefore, be made between the causes of chilling and freezing injuries. Susceptibility to chilling injury and its manifestations vary widely among different commodities. Susceptibility to the disorder means that the lowest safe temperature for these commodities will be well above the lowest non-freezing temperatures.

Table 18 summarises the physical symptoms of chilling injury and the lowest safe storage temperature for some fruits. A common symptom is pitting of the skin, usually due to the collapse of the cells beneath the surface, and the pits are often discoloured. High water loss may occur, which accentuates the extent of pitting. Browning of flesh tissues is also a common feature. Browning often first appears around the vascular (transport) strands in fruit, probably as a result of the action of the enzyme polyphenol oxidase on phenolic compounds released from the vacuole after chilling, though this has not been proved in all cases. Fruit that has been picked immature will fail to ripen or will ripen unevenly or slowly after chilling. Degreening of citrus is slowed by even mild chilling. Water-logging of leafy vegetables and some fruits such as papaya is also often observed. The symptoms of chilling injury normally occur while the produce is at low temperature, but sometimes will only appear when the produce is removed from the chilling temperature to a higher temperature. Deterioration may then be quite rapid, often within a few hours.

Chilling injury causes the release of metabolites, such as amino acids and sugars, and mineral salts from cells which together with the degradation of cell structure provide an excellent substrate for the growth of pathogenic organisms, especially fungi. These are often present as latent infections or may contaminate

Figure 30 Storage life at various temperatures of produce with no (A), slight (B), or high (C) sensitivity to chilling injury. (From Tomkins, R.G. The choice of conditions for the storage of fruits and vegetables. East Malling, England: East Malling Research Station: 1966. Ditton Laboratory Memoir No. 91. With permission.)

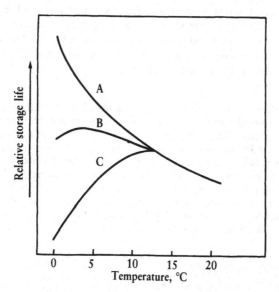

the produce during harvesting, transport and marketing. For this reason increased rotting is a common occurrence in tropical produce after low temperature storage. Another consequence of chilling is the development of off-flavours or odours. The complex array of symptoms suggests that several factors are operative in the development of chilling injury. Thus commodities grown in different areas may behave differently, and varieties of the same crop can also behave quite differently in response to similar temperature conditions.

The temperatures quoted in Table 18 refer to the limiting or critical temperature below which some physical symptom of chilling injury will usually be observed. If the temperature is just below this critical temperature then relatively long exposure to the temperature will be required before injury is observed. Injury will generally appear more quickly and will be more severe the further the temperature is below the critical chilling temperature. Storage of the commodity may be possible for a useful period of time at temperatures slightly below the critical temperature where there is only a slight susceptibility to chilling injury. This relation between storage life at various temperatures and sensitivity to chilling is illustrated in Figure 30. Such a relationship also holds for the development of physiological disorders.

The most obvious method for the control of chilling injury is to determine the critical temperature for its development in a particular fruit and not expose the commodity to temperatures below that critical temperature. But exposure for only a short period to chilling temperatures with subsequent storage at higher temperatures may prevent the development of injury. This has been found effective for preventing black heart in pineapple, woolliness in peach, and flesh browning in plum, but it is not known whether other produce will respond similarly. It has been claimed that modified atmosphere storage can reduce the extent of chilling in some produce and also that maintenance of high relative humidity, both in storage at low temperature and after storage, will minimize pitting. More research is needed to confirm these claims.

Figure 31 Chilling injury in avocado fruit manifested as internal flesh (mesocarp) browning due to the breakdown of cell compartmentation and the action of polyphenol oxidases to produce tannins (top). The whole fruit were stored 21 days at 5°C and then ripened at 20°C for 5 days. The fruit on the bottom was stored (whole) in sealed polyethylene bags so creating a modified atmosphere, typically, 7 per cent carbon dioxide, 3 per cent oxygen and 90 per cent nitrogen, which prevented or delayed the development of symptoms of chilling injury. (From Scott, K.J.; Chaplin, G.R. Reduction of chilling injury in avocados stored in sealed polyethylene bags. Trop. Agric. Trinidad 55: 87–90; 1978, with permission.)

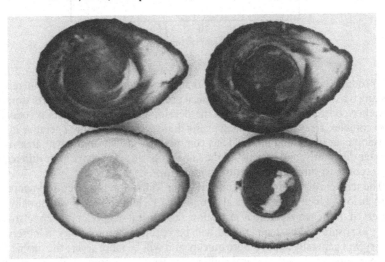

Mechanism of chilling injury

The events leading to chilling injury can be separated into primary events by which the plant cells sense the lowered temperature and the long-term responses or secondary events that ultimately lead to the death of the cells. The primary events are more or less instantaneous and are reversible, at least for a period of time. The secondary events are eventually irreversible and are manifest as the various necrotic and other symptoms of chilling injury. This concept has been developed by J.K. Raison in Australia and J.M. Lyons in the United States and is illustrated in Figure 32.

The critical temperature below which chilling injury will occur is characteristic of the species of plant and the commodity derived from it. Very chilling sensitive plants, such as banana and pineapple have a relatively high critical temperature around 12°C or even higher. Chilling insensitive plants such as apple and pear have a much lower critical temperature around 0°C or below. The critical temperature determines the minimum safe storage temperature of the commodity. Of course, storage below about −1°C is not possible for fresh produce because of freezing damage.

The two most likely causes of chilling sensitivity are: (a) a low temperature induced change in the physical properties of cell membranes due to changes in the physical state of membrane lipids (the membrane lipid hypothesis of J.K. Raison and J.M. Lyons); and (b) low temperature induced dissociation of enzymes and other proteins into their structural sub-units resulting in alteration of the kinetics of enzyme activity and changes in structural, cytoskeletal proteins such as tubulin.

Figure 32   Diagrammatic time sequence of events leading to chilling injury (After G.R. Chaplin, personal communication).

PRIMARY EVENTS	SECONDARY EVENTS		CHILLING INJURY
Reversible	Irreversible		
TIME			
Physical changes in membrane lipids Dissociation of enzymes/proteins	Result in impaired: − ion movements through membranes − metabolism : respiration : photosynthesis : protein synthesis etc. − protoplasmic streaming	membrane breakdown	necrosis, visible symptoms of injury

The lipid hypothesis is supported by data obtained using a number of physical techniques including differential scanning calorimetry (Figure 33), electron spin resonance spectometry and fluorescence polarization of molecular probes intercalated into plant cell membranes. These data showed that there is a change in the physical properties of extracted membrane lipids at characteristic temperatures in the range 7–15°C. The characteristic temperatures were found to coincide with the critical temperatures below which particular plant tissues or fruits show symptoms of chilling injury. Only a very small proportion, <10 per cent, of the membrane lipids undergo the physical change which is probably a phase separation. The particular lipids involved are probably disaturated forms of phosphatidylglycerol (PG) which have high melting points. PG is located in chloroplast membranes. Chilling insensitive plants have relatively low concentrations of these PGs which would account for their critical temperatures for the physical change being around or below 0°C. The lipids responsible for the physical changes in other cellular membranes, eg mitochondria, plasmalemma and tonoplast, have not been identified but it is assumed that a similar mechanism is involved.

A physical change in the membrane lipids with lowering of the temperature would cause changes in the properties of the membranes. For example, ion and metabolite movements would be affected as would activities of membrane-bound enzymes. These could in turn cause imbalances in metabolism with eventual disruption of the various membranes leading to breakdown in cellular compartmentation, death of the cells and the appearance of symptoms of chilling injury.

There is some evidence from both animals and plants that several enzymes of cellular metabolism undergo dissociation at temperatures approaching 0°C. Some multimeric enzymes split into their component sub-units with a consequent loss of enzymic activity and change in some kinetic properties. Some enzymes of both respiratory and photosynthetic metabolism are affected. The consequences of such changes in the relative activities of some enzymes will be imbalances in metabolism which would eventually lead to death of cells. The toxin hypothesis of chilling injury in which toxic products of metabolism, such as acetaldehyde, accumulate could be explained by imbalances in metabolism. Structural proteins of the cell's cytoskeleton, such as tubulin, are cold-labile and undergo dissociation at low temperatures. This could account for the effect of low temperature on protoplasmic streaming which is especially sensitive in chilling sensitive plants.

Further research is required to fully elucidate the mechanisms of chilling injury. If it should turn out that only a few proteins are involved in the synthesis of the key membrane lipids and as cold-labile enzymes of metabolism then it may be possible in future to genetically engineer these proteins to make plants less chilling sensitive and so improve the low temperature storage of sub-tropical and tropical fruits and vegetables.

PHYSIOLOGICAL DISORDERS

Physiological disorders affect mainly deciduous tree fruits, such as apple, pear and stone fruits, and most citrus fruits. Most of these disorders affect discrete areas of tissue (Figure 34). Some disorders may affect the skin of the produce but leave the underlying flesh intact; others affect only certain areas of the flesh or the core region.

With most disorders, the metabolic events leading to a manifestation of the symptoms are not fully understood and with many disorders are not understood at all. The discovery of most disorders could be considered 'non-scientific'. Cool

Figure 33 Thermograms prepared by differential scanning calorimetry of polar lipids from mitochondria of chilling sensitive and chilling insensitive plants. Each sample was scanned 10 C degrees per minute at a sensitivity of 0.2 mcal per second. Chilling sensitive plants with critical temperatures indicated by the thermally induced physical changes around 14 to 15°C: A, cucumber; B, mung bean; C, sweet potato roots. Chilling insensitive plants with no indication of a thermally-induced physical change in the temperature range −15 to 45°C: D, artichoke roots; E, wheat roots; F, barley roots. Sample size was 3.6–11.0 mg of lipid. (From Raison, J.K.; Orr, G.O. Phase transitions in liposomes formed from the polar lipids of mitochondria from chilling-sensitive plants. Plant Physiol. 81: 807–11; 1986. With permission.)

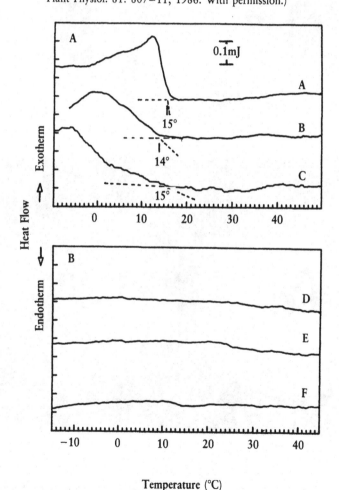

Figure 34 Physiological disorders of apple: top left, superficial scald; top right, Jonathan spot; bottom left, low temperature breakdown; and bottom right, brown heart. (Courtesy K.J. Scott, NSW Department of Agriculture.)

store operators or shipping agents have held fruit at low temperatures and found that they developed a variety of browning conditions. These conditions were given descriptive names as there was no other way of classifying the disorders. These names are still the only classification used. The apple has been studied more intensely than other commodities and also appears to have the greatest variety of physiological disorders. Table 19 lists some of these disorders and their symptoms. These disorders require low temperature storage, usually at less than 5°C, for development of symptoms. Each disorder is, therefore, presumed to be derived by a different metabolic route, although this may not prove to be true when their biochemistry is elucidated. Disorders in other fruits are shown in Table 20. When more research effort is devoted to commodities other than the apple, the list of physiological disorders will undoubtedly increase. There is no reason to believe that apple should be more prone to develop disorders than other commodities.

The early studies on disorders found that, although a particular variety may be susceptible to a certain disorder, not all fruits develop the disorder. Susceptibility to disorders was shown to be dependent on a number of factors, such as maturity at harvest, cultural practices, climate during the growing season, produce size, and harvesting practices. The risk of a fruit developing a particular disorder can, therefore, be minimized by identifying susceptible fruits and not storing them for long periods. But the market often has requirements that result in a preference for the type of fruit that is highly susceptible to a disorder, for example, with the Jonathan apple, the consumer prefers large fruits with intense red colouration,

Table 19 Some physiological disorders of apple

Disorder	Symptoms
Superficial scald	Slightly sunken skin discolouration, may affect whole fruit
Sunburn scald	Brown to black colour on areas damaged by sunlight during growth
Senescent breakdown	Brown, mealy flesh; occurs with overmature, overstored fruit
Low temperature breakdown	Browning in cortex
Soft (or deep) scald	Soft, sunken, brown to black, sharply defined areas on the surface and extending a short distance into the flesh
Jonathan spot	Superficial spotting of lenticels; occurs at higher temperatures
Senescent blotch	Grey superficial blotches on over-stored fruit
Core flush (brown core)	Browing within core line
Water core	Translucent areas in flesh; may brown in storage
Brown heart	Sharply defined brown areas in flesh; may develop cavities

but such fruits are susceptible to low temperature breakdown. Thus methods had to be developed to enable susceptible produce to be stored to meet consumer demand.

Various systems of temperature modulation have been developed to minimize development of some disorders. Lowering the temperature in steps from 3°C down to 0°C in the first month of storage effectively minimizes the development of low temperature breakdown and soft scald in apple. Low temperature breakdown of apple and stone fruits can also be reduced by raising the temperature to about 20°C for a few days in the middle of the storage period and then returning fruit to low temperature. Such methods have not been widely adopted in commercial practice because of the logistical problems of having a whole room of produce ready to treat at one time and the difficulty of rapidly changing the temperature of a room full of fruit. A further problem is that any increase in the storage temperature will increase respiration and thus result in shortening the storage life of produce held in the same room that is not susceptible to the disorder.

Controlled atmosphere storage can completely prevent Jonathan spot when as little as 2 per cent carbon dioxide is present. The incidence of core flush and various forms of flesh breakdown in apple is also often reduced in controlled atmosphere storage, but in some instances the level of breakdown has been reported to be increased in controlled atmosphere storage. This increase has been attributed to factors associated with controlled atmosphere storage other than the composition of the atmosphere. The enclosed room that is required for controlled atmosphere storage results in a high humidity atmosphere, restricted ventilation rates, and the accumulation of fruit volatiles in the atmosphere. These conditions are conducive to the development of apple breakdown. Superficial scald is another disorder that is enhanced in controlled atmosphere storage by

Table 20 Some physiological disorders of fruits other than apple

Produce	Disorder	Symptoms
Pear	Core breakdown	Brown, mushy core on over-stored fruit
	Neck breakdown, vascular breakdown	Brown to black discolouration of vascular tissue connecting stem to core
	Superficial scald	Grey to brown skin speckles; occurs early in storage
	Over-storage scald	Brown areas on skin in over-stored fruit
	Brown heart	Same as for apple
Grape	Storage scald	Brown skin discolouration of white grape varieties
Citrus	Storage spot	Brown sunken spots on surfaces
	Cold scald	Superficial grey to brown patches
	Flavocellosis	Bleaching of rind; susceptible to fungal attack
	Stem-end browning	Browning of shrivelled areas around stem-end
Peach	Woolliness	Red to brown, dry areas in flesh
Plum	Cold storage breakdown	Brown, gelatinous areas on skin and flesh

these conditions (see below). Controlled atmosphere storage can also create new disorders if the produce is exposed to very high levels of carbon dioxide or low levels of oxygen for prolonged periods. The critical level of carbon dioxide that induces brown heart of apple and pear varies among different varieties and may be as low as 1 per cent. Low oxygen injury is characterized by the development of alcoholic off-flavours, produced by anaerobic metabolism, in addition to browning of the tissue.

The ultimate method for the prevention of a disorder is to understand the metabolic sequence that leads to the development of the disorder and then prevent that metabolism from occurring. Chemical control is an obvious measure to prevent the development of disorders, but it is not necessarily the sole method possible. Storage disorders may also be minimized by physical and cultural treatments and by breeding less susceptible cultivars.

Skin blemishes are generally the more serious problem as even quite small skin marks render the fruit unacceptable in many markets. Internal defects can be tolerated to a greater extent as the consumer buys on visual inspection external appearance and even upon consumption may never be aware of a small amount of internal browning. The skin disorders, bitter pit and superficial scald of apple, have received considerable attention, and control measures have been developed for both disorders (see Chapter 10).

Most is known about the metabolism of superficial scald. Early studies (before 1930) led to the hypothesis that superficial scald was caused by a toxic, volatile organic compound that accumulated in the apple during cool storage. In the 1960s, workers in Australia isolated α-farnesene, a C15 sesquiterpene hydrocarbon from susceptible apple varieties, and suggested that it was the precursor of superficial scald. Being a fat-soluble compound, it accumulates in the lipid fraction on the skin. Oxidation products of α-farnesene have been claimed to lead to the collapse of the cells and to tissue browning. Control of the disorder is achieved commercially by the application of various synthetic antioxidants, such as diphenylamine and ethoxyquin, which protect α-farnesene against oxidation. Chilling injury is usually regarded as developing through a different metabolic route to that for superficial scald, but α-farnesene has been shown to accumulate in banana during the development of chilling symptoms, suggesting that there may be a metabolic similarity in the two disorders.

If satisfactory control methods are not available, the ultimate method of avoiding any physiological disorder is to hold susceptible fruits at a temperature high enough to avoid the disorder being a problem. This temperature is usually 3 to 5°C, but is sometimes greater than 5°C. This partially negates the idea of using low temperature to minimize respiration, but it is preferable to market overmature produce than to have a disfiguring disorder present.

MINERAL DEFICIENCY DISORDERS

Fruit and vegetables often show various browning symptoms that have been attributed to deficiencies in some mineral constituents of the produce. These disorders are prevented by the addition of the specified mineral either during growth or postharvest, although for most disorders the actual role of the mineral in preventing the disorder has not been established. Plants require a balanced mineral intake for proper development, so a deficiency in any essential mineral will lead to maldevelopment of the plant as a whole. It can be said that the condition is a physiological disorder if the fruiting organ or actual 'vegetable' portion is affected rather than the whole plant.

Calcium has been associated with more deficiency disorders than other minerals, and some examples are shown in Table 21. Some of these disorders, such as blossom-end rot of tomatoes, can be readily eliminated by the application of calcium salts as a preharvest spray while for others, such as bitter pit of apples, only partial control is obtained. However, this variability in degree of control is probably related to the amount of calcium taken up by the fruit. For example, use of postharvest dipping at sub-atmospheric pressures, which markedly increases the uptake of calcium, usually results in total elimination of bitter pit.

A substantial amount of the added calcium binds with pectic substances in the middle lamella and with membranes generally and may prevent disorders by

Table 21   Calcium-related disorders of fruit and vegetables

Produce	Disorder
Apple	Bitter pit, lenticel blotch, cork spot, lenticel breakdown, cracking, low temperature breakdown, internal breakdown, senescent breakdown, Jonathan spot and water core
Avocado	End spot
Bean	Hypocotyl necrosis
Brussels sprout	Internal browning
Cabbage	Internal tipburn
Chinese cabbage	Internal tipburn
Carrot	Cavity spot, cracking
Celery	Blackheart
Cherry	Cracking
Chicory	Blackheart, tipburn
Escarole	Brownheart, tipburn
Lettuce	Tipburn
Mango	Soft nose
Parsnip	Cavity spot
Pear	Cork spot
Pepper	Blossom-end rot
Potato	Sprout failure, tipburn
Strawberry	Leaf tipburn
Tomato	Blossom-end rot, blackseed, cracking
Watermelon	Blossom-end rot

stengthening structural components of the cell without alleviating the original causes of the disorder. The strengthening of cell components may prevent or delay the loss of cell compartmentation and the enzyme reactions that cause browning symptoms. Calcium has been found to be relocated in apples during storage which raises the possibility that a local deficiency can be created in one part of the tissue during storage resulting in a manifestation of a physiological disorder in that region.

Calcium has been shown to affect the activity of many enzyme systems and metabolic sequences in plant tissues. The addition of calcium to intact fruit or fruit slices generally suppresses respiration but the response is concentration dependent. The activities of isolated pectic enzymes, pectinmethylesterase (PME), exopolygalacturonase (exo PG) and endopolygalacturonase (endo PG), have shown differential responses to calcium concentration. The activity of PME is initially increased by increasing concentrations of calciums but is inhibited at higher concentrations. The large form of endo PG (PG1) extracted from tomato fruit is slightly stimulated by concentrations of calcium that inhibit the smaller endo PG forms of the enzyme. Calcium is needed for the activity of exo PG, kinases and a range of other enzymes. The ability of calcium to regulate these various systems has led to speculation that calcium may have a role in the initiation of the normal fruit ripening process. It is also possible that calcium prevents or delays the appearance of some physiological disorders by maintaining normal metabolism.

OTHER MINERALS

Boron deficiency in apple leads to a condition known as internal cork. This condition is marked by pitting of the flesh and is often indistinguishable from bitter pit. The differences between the two disorders are that internal cork is prevented by the application of boron sprays, bitter pit responds to calcium treatment, and the former develops only on the tree while the latter can develop after harvest.

The major mineral in plants is potassium (K), and both high and low levels of potassium have been associated with abnormal metabolism. High potassium has been associated with the development of bitter pit in apple so that both high potassium and low calcium are correlated with pit development. Low potassium is associated with changes in the ripening tomato and delays the development of full red colour by inhibiting lycopene biosynthesis.

There may be roles for other minerals in the development of other disorders. Injections of copper, iron and cobalt have induced symptoms similar to low temperature breakdown and superficial scald in apple, but this does not necessarily mean they have a role in the development of the natural disorder. Heavy metals, especially copper, act as catalysts for the enzymic systems which lead to enzymic browning, the browning of cut or damaged tissues that are exposed to air. The levels of these metals are important in processed fruit and vegetables, whether they are derived from the produce or from metal impurities that are included during processing.

FURTHER READING

Faust, M.; Shear, C.B.; Williams, M.W. Disorders of carbohydrate metabolism of apples. (watercore, internal breakdown, low temperature and carbon dioxide injuries). Bot. Rev. 35: 168–94; 1969.

Ferguson, I.B. Calcium in plant senescence and fruit ripening. Plant Cell Environ. 7: 477–89; 1984.

Fidler, J.C.: Wilkinson, B.G.: Edney, K.L.; Sharples, R.O. The biology of apple and pear storage. Slough, England: Commonwealth Agricultural Bureaux: 1973.

Graham, D.; Patterson, B.D. Responses of plants to low, nonfreezing temperatures: proteins, metabolism, and acclimation. Annu. Rev. Plant Physiol. 33: 347–72; 1982.

Hall, E.G.: Scott, K.J. Storage and market diseases of fruit. Melbourne, Australia; CSIRO; 1977.

Lallu, N. Postharvest aspects of Asian pears. II. Storage. Asian pear series. Publication No. 8. Wellington: N.Z. Apple and Pear Marketing Board,Technical Bull. No. 9; 1985.

Leshem, Y.Y. Membrane phospholipid catabolism and Ca^{2+} activity in control of senescence. Physiol. Plant. 69: 551–9; 1987.

Lyons, J.M. Chilling injury in plants. Annu. Rev. Plant Physiol. 24: 445–66; 1973.

Patterson, B.D.; Graham, D. Temperature and metabolism. Davies, D.D. ed. The biochemistry of plants. Vol. 12. London: Academic Press; 1987: 153–99.

Pierson, C.F.; Ceponis, M.J.; McCulloch, L.P. Market diseases of apples, pears and quinces. Washington, DC: US Department of Agriculture; 1971. Agriculture Handbook No. 376.

Poovaiah, B.W. Role of calcium and calmodulin in plant growth and development. HortScience 20: 347–52; 1985.

Poovaiah, B.W. Role of calcium in prolonging storage life of fruits and vegetables. Food Technol. 40 (5) 86–9; 1986.

Raison, J.K. Alterations in the physical properties and thermal responses of membrane lipids: Correlations with acclimation to chilling and high temperature. St. John, J.B.; Berlin, E.; Jackson, P.C., eds. Frontiers of membrane research in agriculture. Beltsville Symposium 9. Totowa: Rowman & Allanheld; 1985: 383–401.

Raison, J.K.; Lyons, J.M. Chilling injury: a plea for uniform terminology. Plant Cell Environment. 9: 685–6; 1986.

Shear, C.B. Calcium-related disorders of fruits and vegetables. HortScience 10: 361–5; 1975.

8
Quality evaluation of fruit and vegetables

The term quality defies complete and objective definition, and for the consumer it is largely a subjective judgement. The answer with respect to fruit and vegetables will vary between commodities and, for a particular commodity, will also depend on the position of the recipient in the distribution chain. As an example: what is a good quality pear?

1. To the producer a good quality pear is one that secures a maximum price in the market at a particular time of the season. He must judge between producing either a high quality product by intensive care but with, usually, a lower yield, or a poorer quality product by less intensive care but with a higher yield. For example, a higher total return may be gained for early-picked pear, despite the fact that the quality may not be ideal for the consumer, and the yield on a weight basis will be generally lower than a later pick of the crop.
2. To the shipper, it is a hard green pear. It is one that is capable of being transferred from the orchard to the market without damage—the harder the better.
3. To the canner, good quality is a ripe, but firm pear. The consumer requires a canned pear to be soft, but the pear needs to be firm enough to retain its shape undamaged during processing and subsequent market handling.
4. To the consumer of the fresh fruit it is a soft, ripe product. It must melt in the mouth and be juicy. Skin colour is also often important, whereas to the canner it is immaterial as the skin is removed prior to processing.

Quality may, therefore, be defined in terms of end use. In these terms, produce quality requirements which are commonly encountered refer to market, storage, transport, eating and processing qualities. The marketing of fresh fruit and vegetables is aimed eventually at appealing to the consumer for whom tradition plays a major role in determining the acceptability of a food item. The eating habits of man are conservative.

QUALITY STANDARDS

Many countries, especially those exporting fruit and vegetables, have established quality standards to ensure that the buyer can rely on a certain minimum standard. A wide variety of quality determinants is employed, including size, colour, maturity and extent of blemishes. These standards are usually enforced by government authorities through an inspection service. The increase of direct marketing from grower to supermarket chains is leading to the imposition of standards to predetermined specifications by the supermarkets.

There is no universal set of quality standards for any given commodity. Each country has its own criteria depending on local circumstances. Different standards may apply for produce for home consumption and for export. Generally only the higher quality lines are exported, because of the longer time the produce has to survive before consumption.

QUALITY CRITERIA

Important factors in quality for the consumer are:

1. appearance, including size, colour, shape;
2. condition and absence of defects;
3. texture;
4. flavour; and
5. nutritional value.

Appearance

This is probably the most important quality factor determining the market value of produce, as people 'buy with their eyes'. They have learnt from past experience to associate desirable qualities with a certain external appearance. A rapid, visual assessment can, with experience, be made on the criteria of size, shape, colour, condition (such as freshness), and the presence of defects or blemishes.

Size is an important criterion of quality which can be easily measured either by circumference, diameter, length, width, weight or volume. Many fruits are graded according to size, usually by diameter measurement, with similar sizes of fruit being packed together to give a uniform pack, which facilitates marketing and retail sales. For example, certain size standards are adhered to for export apples based on fruit diameter or count number per package. These standards are specific to a particular package, and also depend on their export destination. The maximum size limits for apple ensures that large apples, which are highly susceptible to postharvest physiological disorders such as storage breakdown, are not exported. Particular packing arrays and individual wrapping of fruit, especially the top layer, may also be mandatory or recommended. The packing array will determine the number of fruits per package for some commodities. An example of a commodity graded by length and diameter is the carrot. For many other commodities weight is the standard determinant.

Shape is a criterion which often distinguishes particular cultivars of fruit. Characteristic shapes are usually demanded by the consumer who will often reject a commodity which lacks the characteristic shape. For example, attempts to market a straight banana were unsuccessful, apparently because this was considered abnormal. Shape is especially important in apple cultivars, a premium price being obtained for fruit with a well-developed shape characteristic of the particular cultivar. Misshapen fruit and vegetables are poorly accepted. Any deviation in shape usually brings a lower price. Shape is often a problem in breeding programs. Although a superior eating or storing product may be obtained by breeding a new cultivar, if its shape is unusual it will be less readily accepted in the market and will require extensive re-education of the consumer through advertising programs.

One of the distinguishing features of fruit and vegetables is that they are the only major group of natural foods with a variety of bright colours. They are often used merely to brighten up the presentation of foods. Parsley contains relatively high levels of ascorbic acid, carotene, thiamin, riboflavin, iron and calcium compared to fruit and other vegetables, yet is used almost exclusively in Western society to add colour and flavour to meat and fish dishes. A bright red apple is most valued in some markets, yet the colour of the skin adds nothing to its eating quality or to its nutritive value. Similarly, the red colour (lycopene) of tomato fruit is not nutritionally significant but is a criterion of ripeness. Colour changes in ripening fruit have been correlated by the consumer with the conversion of starch to sugar, that is, sweetening, and the development of other desirable attributes so that the correct skin colour is often all that is required for a decision to purchase the commodity. Such subjective assessments may be misleading. For example, if fruits such as the banana are ripened at higher than optimum temperatures, full loss of green colour does not occur, and consumers, particularly those from temperate climates, would show some degree of buyer resistance even though the flesh is adequately ripened. Standardized colour charts are used in the visual assessment of ripeness in many fruits, for example, tomato, pear, apple and banana.

Condition and defects

Condition is a quality attribute referring usually to freshness and stage of senescence or ripeness of a commodity. Wilted leafy vegetables obviously lack condition and are generally unacceptable to the consumer. Similarly, shrivelled fruit, due to excessive loss of moisture, also lack condition. The loss of condition is most marked towards the end of the shelf-life of a commodity in the retail outlet. Prevention of such loss of condition can be achieved by improved storage conditions, which usually equates with cool storage but may be as simple as shading the produce, especially leafy vegetables, from direct sunlight.

Skin blemishes, such as bruises, scratch marks and cuts, detract from appearance and on markets in Western countries detract from market price, even when the blemishes reduce neither keeping quality nor eating quality. In developing countries a premium price may be obtained for produce that is free from blemishes, but there will still be a market for lower grade produce which would possibly be unsaleable in Western countries. Consequently a greater proportion of a crop is likely to be sold and consumed in developing countries.

Not all consumer correlations are necessarily based on valid scientific evidence. Some commodities such as banana frequently develop latent infections of the skin during ripening as a consequence of the weakening of the skin structure so that the ripe product is often flecked, that is, yellow with 'black spots'. If this mould growth is controlled, as is now possible, the product then may be regarded by some consumers with suspicion—'It doesn't look right!'. 'Normal' appearance is extremely important in the market place. Acceptable appearance, however, may differ between countries and between different regions within a country. In Japan great importance is attached to the unblemished appearance of the fruit, for example, a perfectly even, yellow colour is required in banana peel.

Thus appearance is a major determinant of quality, especially because it is often the only criterion available to the buyer of the commodity. On the spot taste-testing is rarely practised or encouraged at the retail level, although it is more general at the commercial markets from which the retailers purchase their commodities.

Texture and flavour

Texture is the overall assessment of the feeling the food gives in the mouth. It is a combination of sensations derived from the lips, tongue, walls of the mouth, teeth and even the ears. Each area is sensitive to small pressure differences and responds to different attributes of the produce. The lips assess the type of surface being presented so that they can distinguish between hairy and smooth surfaces. The teeth determine rigidity of structure. They are sensitive to the amount of pressure required to cleave the food and the manner in which the food gives way under the applied force. The tongue and walls of the mouth are sensitive to the types of particles generated following cleavage by the teeth—whether they are soft and mushy or discrete lumps. The extent of expressed juice is also assessed. The ears determine the sounds of the food being chewed. This is important in foods, such as celery, apple and lettuce, where crispness is a desirable attribute. The cumulative effect of these responses creates an overall impression of the texture of the produce.

Flavour comprises two factors: taste and aroma. Taste is due to the sensations felt on the tongue. The four main taste sensations are sweet, salt, acid (sour) and bitter. Each sensation is largely perceived at a specific area of the tongue (Figure 35). All food tastes elicit a response on one or more areas. The taste of fruit and vegetables is usually a blend or balance of sweet and sour, often with overtones of bitterness due to tannins. They are not naturally salty. Aroma is due to stimulation of the olfactory senses with volatile organic compounds, some of which have been outlined previously in Chapter 2.

Nutritional value

Nutrition is probably the least important consideration in determining whether a consumer purchases a commodity, since most essential nutrients can neither

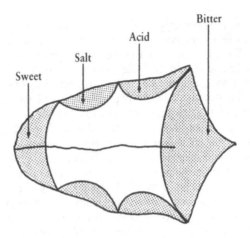

Figure 35   Areas of taste sensations on the tongue.

be seen nor tasted. The main nutrient of value is vitamin C, and fruit and vegetables are the sole source of vitamin C in the diet of many people. Few people, however, would decide to buy a particular piece of fruit, because it had more vitamin C than another type of fruit (see Table 4 for vitamin C contents).

Improved nutritive value should be aimed at by all connected with any aspect of the fruit and vegetable industry, as it is a means of upgrading the health of the community without changing their food habits. But where varieties with superior nutritional value have been produced, acceptance by growers and consumers has been marginal. Some type of government subsidy or coercion is imperative if nutritional gains are to be made within a reasonable time. This was well demonstrated with corn (maize). Breeding programs have developed corn varieties with high levels of the amino acids, lysine and tryptophan, thus overcoming a serious nutritional deficiency in normal corn varieties. Acceptance, and therefore the benefit, has been limited to date. The new varieties looked different and had some different cultural requirements so that both farmers and consumers saw no benefit in the new corn. Population groups with impoverished diets tend to be poorly educated; therefore, it is difficult to convince them of benefits to be gained from a change in diet.

POSTHARVEST FACTORS INFLUENCING QUALITY

Not all the changes in harvested produce cause loss of quality. Some changes that occur after harvest are essential for the attainment of the desired degree of eating quality. Many tropical and subtropical fruits, such as banana, mango and tomato, are picked at the mature-green stage of development and then allowed to ripen off the plant to attain optimum eating quality. This is essential for avocado, which will not ripen while attached to the tree. The main concern with other postharvest produce, however, is generally with preventing deterioration of the existing quality. Loss in quality can be caused by a variety of means which may be grouped under our main headings: metabolic factors, transpiration, mechanical injury and microorganisms.

Metabolic factors include normal senescence or abnormal metabolism leading to the development of physiological disorders. The loss from physiological disorders is more spectacular than from general senescence effects, but is less of a problem in the overall storage of fruit and vegetables. Transpiration (loss of water) can result in rapid loss in quality. Severe wilting of leafy vegetables can be induced by storage for less than one day under hot dry conditions. Although wilting mainly affects texture, consumers strongly resist purchasing wilted produce. Mechanical injury causes loss of visual quality with unsightly marks. Injuries lead to an increase in general metabolism as the produce tries to seal off the demaged tissues. Furthermore, transpiration increases as the natural barriers against loss of water have been damaged. The microorganisms of interest are mainly moulds. Where favourable conditions for their growth are present, such as optimum temperature, pH and humidity conditions, the growth of moulds can be rapid and can cause extensive losses. Some of the major factors which can contribute to loss of quality after harvest are discussed below.

Harvesting
Mechanical damage during harvesting and subsequent handling operations can result in defects on the produce and permit invasion by disease-causing microorganisms. The inclusion of dirt from the field can aggravate this situation. Produce can overheat and rapidly deteriorate during temporary field storage. Failure to sort and discard immature, overripe, undersized, misshapen, blemished, or otherwise damaged produce creates problems in the subsequent marketing of the produce.

Transport and handling
Rough handling and transport over bumpy roads damages produce by mechanical action. At high temperatures the produce will become overheated, especially if there is inadequate cooling or ventilation. Transport on open trucks can result in sun-scorch of the exposed produce; loads should be protected from the sun. Severe water loss, especially from leafy vegetables, can also occur under these conditions. Inappropriate packaging may cause physical damage of produce due to bruising or abrasion as the commodity moves about during transport.

Storage
Delay after harvest in placing produce in cool storage often causes rapid deterioration in quality. Poor control of storage conditions, over-long storage and inappropriate storage conditions for a particular commodity will also result in a poor quality product. With mixed storage of different commodities, ethylene

produced from ripening fruit can promote rapid senescence of leafy vegetables. Storage at temperatures that are too low may induce physiological disorders or chilling injury.

Marketing

A serious reduction in quality can occur in produce displayed for lengthy periods in retail outlets because of poor organization of marketing. The current practice of displaying cleaned potatoes in relatively bright light in supermarkets leads to greening and production of the toxic glycoalkaloid, solanine.

Many of the above practices are aggravated by pathological problems resulting from rotting by fungi or bacteria. High humidity conditions are conducive to microbial growth, especially in produce damaged during harvesting and handling.

Residues of pesticides and other chemicals are a further factor in postharvest quality. These are often applied preharvest in the form of insecticides or herbicides. Fungicides may be used both preharvest and postharvest to prevent rotting, and fumigants may be used for insect disinfestation, especially in export trade. All of these chemucals can leave residues in the commodity and, although usually not detectable by the consumer, must be considered in relation to the health risks to the community.

DETERMINATION OF MATURITY

There is a clear distinction between 'physiological maturity' and 'commercial maturity'. The former is a particular stage in the life of a plant organ, and the latter is concerned with the time of harvest as related to a particular end-use that can be translated into market requirements.

Physiological maturity refers to the stage in the development of the fruit or vegetable when maximum growth and maturation has occurred. It is usually associated with full ripening in a fruit. The physiologically mature stage is followed by senescence. Clear distinction between the three stages of development, namely, growth, maturation and senescence, in the development of a plant organ is not always easy, since the transitions between the stages are often quite slow and indistinct. But measurement of respiration and ethylene production and various chemical determinations, such as sugar acid ratios, can give reliable estimates of the stage of maturity of specific commodities (see Chapter 3).

Commercial maturity is the state of a plant organ required by a market. Commercial maturity commonly bears little relation to physiological maturity and may occur at any stage during development or senescence. The terms immaturity, optimum maturity and over-maturity relate to these requirements. There must be an understanding of each of them in physiological terms, particularly where storage life and quality when ripe are concerned. Some examples of commercial maturity in relation to physiological age are shown in Figure 36.

Ripening, as applied to fruit, is the process by which a fruit develops its maximum desired eating quality. The potential quality is determined by many factors, of which the stage at which the fruit was harvested is most important. A fruit is ripe when it has developed the maximum desired eating quality. Before this stage

Figure 36 Commercial maturity in relation to developmental stages of the plant. (Adapted from Watada, A.E.; Herner, R.C.; Kader, A.A.; Romani, R.J.; Staby, G.L. Terminology for the description of developmental stages of commercial crops. HortScience. 19: 20–1; 1984.)

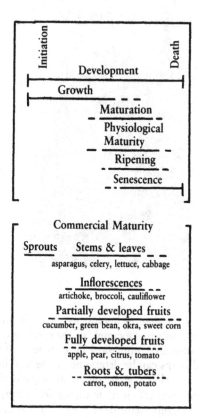

it is underripe and after it is overripe. These stages of ripeness cannot be clearly defined physiologically but are subjective judgements and will vary among consumers.

DETERMINATION OF COMMERCIAL MATURITY

The index of commerical maturity usually involves some expression of the stage of development or maturation and requires measurement of some characteristic known to change as the fruit or vegetable matures. It may involve making decisions about levels of market and consumer acceptability. It may also involve objective measurements or subjective judgements or both.

Figure 37   Pea maturometer.

The determination of the time of harvest for peas for processing by canning or freezing provides an interesting example of the need for close control of commercial maturity. An objective, numerical measurement of maturity of shelled peas, beans and broad beans can be determined by several instruments specially designed for this purpose. They operate on the principle of the force required to shear the seed. These instruments include the tenderometer, used mainly in the USA and UK, and the maturometer used mainly in Australia (Figure 37). The end result of the research and development needed to overcome the problems in determining optimum maturity of shelled vegetables for processing has been the provision for the consumer of a generally uniform, high quality, canned and frozen product. Such objective standards cannot always be set for maturity determination in many commodities. In the final analysis, judging the time of harvest is often determined by the grower's experience with his own crop in terms of calendar date and various subjective judgements in relation to market requirements.

Many criteria for judging maturity have been used or suggested and include: skin or flesh colour; flesh firmness; electrical or light transmittance characteristics; chemical composition; size and shape; respiratory behaviour or time to ripen; time from flowering or planting (calendar date); heat units. There is considerable grower experience and knowledge, particularly with lesser known tropical fruits, which should be sought out and translated into generally applicable tests. To be practical, maturity tests should be simple and rapid and readily carried out in the field. If they can be non-destructive so much the better.

Colour

The loss of green colour (often referred to as the 'ground colour', that is, the background colour) of many fruits is a valuable guide to maturity. There is initially a gradual loss of intensity of colour from deep green to lighter green and, with many commodites, complete loss of green with the development of yellow, red or purple pigments. Ground colour, as measured by prepared colour charts, is a useful index of maturity for apple, pear and stone fruits, but is not entirely reliable as it is influenced by factors other than maturity.

For some fruits, as they mature on the tree, development of blush colour, that is, additional colour superimposed on the ground colour, can be a good indicator of maturity. Examples are the red or red-streaked apple cultivars and red blush on some cultivars of peach. Such colour development is usually dependent on exposure to sunlight. For other fruits, such as the characteristically green apple cultivar, 'Granny Smith', over-exposure to sunlight leads to an undesirable yellow sunburn, which diminishes the market value of this apple. For certain tropical fruits such as papaya, skin colour is a reliable guide to commercial maturity. The appearance of a trace of yellow at the apical end of papaya may be used to determine time of harvest. Such fruit, however, would benefit from a longer period on the tree, till about one-third of the fruit is yellow, after which ripening is quicker and eating quality is better though shelf-life will be shorter. Thus the grower must balance the improved benefits of better eating quality against the greater risk of losses from overripeness and consequent commercial losses in marketing. In grading the maturity of tomato for the fresh, table market, both the loss of green and the development of red colour are used in the construction of a standard chart. In contrast, external colour can be useless as a guide to maturity in tropical citrus, which remain green, and in pineapple, the flesh of which may be mature though the shell is green.

Objective measurement of colour is possible using a variety of reflectance or light transmittance spectrophotometers. The Hunter Color and Color Difference Meter is an example of the former which is widely used in research work. The US Department of Agriculture has developed several spectrophotometric devices which are gaining some commercial acceptance. The high capital cost of such instrumentation usually restricts them to packing houses with a high throughput of produce. But as labour costs increase it may be expected that such mechanization will increase, for example, electronic colour sorting is now common for lemons and for processing tomatoes.

Flesh firmness

As fruit mature and ripen they soften by dissolution of the middle lamella of the cell walls. This softening can be estimated subjectively by finger or thumb pressure, but a more precise objective measurement, giving a numerical expression of flesh firmness, is possible with a fruit pressure tester or penetrometer. Two commonly used pressure testers, by which the resistance of flesh to the penetration of a

Figure 38   Effegi (bottom) and Magness-Taylor (top) fruit pressure testers.

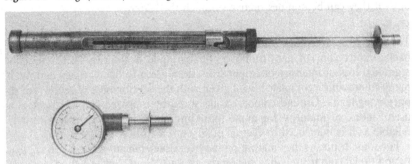

standard plunger is measured, are the Magness-Taylor and UC Fruit Firmness testers, and the smaller and more convenient Effegi instrument (Figure 38). A more elaborate, but not necessarily more effective, test uses machines such as the Instron Universal Testing Instrument. The latter type are often of value only for experimental studies. These three instruments do not necessarily give the same numerical value if used on the same produce, although each instrument will give a reproducible value. It is, therefore, necessary to specify the instrument when reporting pressure test values or attempting to set standards.

Electrical characteristics
Changes in resistance or capacitance as a result of changes in the concentrations of dissolved electrolytes of the flesh during maturation have been demonstrated. These measurements can be useful in the laboratory but appear to be of little practical value.

Chemical measurements
Measurement of chemical characteristics of produce is an obvious approach to the problem of maturity determination, particularly as they can often be related to the palatability of produce. Some useful characteristics are given as examples.

The conversion of starch to sugar during maturation is a simple test for the maturity of some apple varieties. It is based on the reaction between starch and iodine to produce a blue or purple colour. The intensity of the colour indicates the amount of starch remaining in the fruit. This test can also demonstrate the disappearance of starch from the pulp of ripening banana.

Sugar content can be measured directly by chemical means but, since sugar is usually the major component of soluble solids, it is much easier and just as useful to measure total soluble solids in extracted juice with a refractometer or hydrometer. Maturity standards for melons, grape and citrus are often based on the level of soluble solids.

Changes in dry matter can be used for some products in which there is a large increase in the amount of starch or sugar as the fruit mature. Dry matter can be

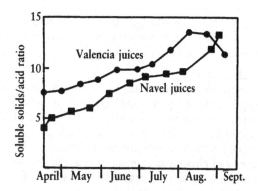

Figure 39 Variation in soluble solids acid ratio in juice from Washington Navel and Valencia Late grown on trifoliata orange rootstock. ■ Washington Navel ● Valencia Late. (Adapted from Chandler, B.V.; Nicol, K.J.; von Biedermann, C. Factors controlling the accumulation of limonin and soluble constituents within orange fruits. J. Sci. Food Agric, 27: 866–76; 1976.)

conveniently and rapidly determined using a microwave oven to dry the material before weighing.

Acidity (titratable acidity) is readily determined on a sample of extracted juice by titration. The loss of acidity during maturation and ripening is often rapid. The sugar acid or total soluble solids acid ratio is often better related to palatability of the fruit than either sugar or acid levels alone. Maturity standards for citrus fruits are commonly expressed as total soluble solids acid ratios, both measurements being on a weight for weight basis. Figure 39 shows the increase in this ratio as the fruit matures on the tree for the two orange cultivars, Valencia and Washington Navel. A minimum soluble solids content is often included in specifications of fruits for processing, commonly for citrus and pineapple, and for grapes for drying or juice production.

The titratable acidity and pH are not directly related as the latter depends on the concentration of free hydrogen ions and buffering capacity of the extracted juice. But pH is a convenient measure, easily made with an inexpensive pH meter, and is widely used, for example, in the wine industry to measure the quality of expressed grape juice.

Fruit shape and size
Fruit shape may be used in some instances to decide maturity. For example, some cultivars of banana become less angular in cross section as development and maturation progress. The fullness of the 'cheeks' adjacent to the pedicel may be used as a guide to maturity of mango and some stone fruits.

Size is generally of limited value as a maturity index in fruit though it is widely used for many vegetables, especially those marketed early in their development, for example, zucchini. With these produce, size is often specified as a quality standard, with large size generally indicating commercial over-maturity and under-sized produce indicating an immature state. This assumption, however, is not always a reliable guide for all produce.

Respiratory behaviour

Commerical maturity can be related to the rise in respiration in climacteric fruits. For longest storage, apple and pear should generally be picked just before this rise, but not too early or quality when ripe will be poor. In practice, the appropriate point on the respiration curve must be related to some characteristic which can be readily determined in the field. The concept of 'green-life' has been developed to indicate the degree of physiological immaturity at the time of harvest and also as a useful expression of potential postharvest life. Green-life is the time from harvest of the fruit to onset of ripening.

Calendar date

For perennial fruit crops grown in seasonal climates which are more or less uniform from year to year, calendar date for harvest is a reliable guide to commercial maturity. This approach relies on a reproducible date for the time of flowering and a relatively constant growth period from flowering through to maturity. Time of flowering is largely dependent on temperature, and the variation in number of days from flowering to harvest can be calculated for some commodities by use of the degree-day concept; such harvesting criteria are often the product of the grower's experience with his crop in his particular environment.

Heat units

An objective measure of the time required for the development of the fruit to maturity after flowering can be made by measuring the degree days or heat units in a particular environment. It has been found that a characteristic number of 'heat units' or 'degree days' is required to mature a crop. Under unusually warm conditions maturity will be advanced and under cooler conditions delayed. The number of degree days to maturity is determined over a period of several years by obtaining the algebraic sum of the differences, plus or minus, between the daily mean temperatures and a fixed base temperature (commonly the minimum temperature at which growth occurs). The average or characteristic number of degree days is then used to forecast the probable date of maturity for the current year and as maturity approaches, it can be checked by other means. This heat unit approach is helpful in planning planting, harvesting, and factory programs for crops, such as corn, peas and tomato, for processing.

FURTHER READING

Abbott, J.A.; Watada, A.E.; Massie, D.R. Effe-gi, Magness-Taylor, and Instron fruit pressure testing devices for apples, peaches, and nectarines. J.Am. Soc. Hortic. Sci. 101: 698–700; 1976.

Anon. Vegetable products, processed. Solids (total) in processed tomato products. Microwave oven drying method. J. Assoc. Off. Anal. Chem. 68: 390–1; 1985.

Arthey, V.D. Quality of horticultural products. London: Butterworths; 1975.

Bourne, M.C. Food texture and viscosity: concept and measurement. New York: Academic Press; 1982.

Gacula, M.C.; Singh, J. Statistical methods in food and consumer research. Orlando: Academic Press; 1984.

Haard, N.F.; Salunkhe, D.K., eds. Symposium: postharvest biology and handling of fruits and vegetables. Westport, CT: AVI; 1975.

Hultin, H.O.; Milner, M., eds. Postharvest biology and biotechnology. Westport. CT: Food and Nutrition Press, Inc.; 1978.

Kader, A.A. et al. Postharvest technology of horticultural crops, Berkeley, CA: University of California; 1985. Special Publication 3311.

Kapelis, J.G. ed. Objective methods in food quality assessment. Boca Raton FL: CRC Press; 1987.

Kramer, A.; Twigg, B.A. Quality control for the food industry. Vol. 1. Fundamentals. 3d ed. Westport. CT: AVI; 1970.

Organisation for Economic Co-operation and Development. International standardisation of fruit and vegetables. Apples and pears, tomatoes, citrus fruit, shelling peas, beans and carrots. Paris; 1976.

Scott, K.J.; Spraggon, S.A.; McBride, R.L. Two new maturity tests for kiwifruit. CSIRO Food Res. Q. 46: 25–31; 1986.

Stone, H.; Sidel, J.L. Sensory evaluation practices. Orlando: Academic Press; 1985.

9
Pathology

Wastage of fruit and vegetables by microorganisms during movement from harvest to consumption can be rapid and severe, particularly in tropical areas where high temperatures and high humidity favour rapid microbial growth. Furthermore, ethylene produced by rotting produce can cause premature ripening and senescence of other produce in the same storage and transport environment, and sound produce can be contaminated by rotting produce. Apart from actual losses due to wastage, further economic loss occurs if the market requirements necessitate sorting and repacking of partially contaminated consignments.

MICROORGANISMS CAUSING POSTHARVEST WASTAGE

Many bacteria and fungi can cause the postharvest decay of fruit and vegetables. However, it is well established that the major postharvest losses of fruit and vegetables are caused by species of the fungi *Alternaria, Botrytis, Diplodia, Monilinia, Penicillium, Phomopsis, Rhizopus* and *Sclerotinia* and of the bacteria *Erwinia* and *Pseudomonas* (Table 22). Most of these organisms are weak pathogens in that they can only invade damaged produce; a few, such as *Colletotrichum*, are able to penetrate the skin of healthy produce. Often the relationship between the host (fruit or vegetable) and the pathogen is reasonably specific, for example, *Penicillium digitatum* rots only citrus and *P. expansum* rots apple and pear, but not citrus. Details of these host–pathogen relationships have been published for a wide range of produce by organizations such as the Department of Agriculture in the USA and CSIRO in Australia. Complete loss of the commodity occurs when one, or a few, pathogens invade and break down the tissues; this initial attack is rapidly followed by a broad spectrum of weak pathogens which magnify the damage caused by the primary pathogens. The appearance of many commodities may be marred by surface lesions caused by pathogenic organisms without the internal tissues being affected.

Table 22   Examples of major postharvest diseases of fresh fruits and vegetables[1]

Crop	Disease	Pathogens
Apple, pear	Lenticel rot	*Phlyctaena vagabunda* Desm. (= *Gloeosporium album* Osterw.)
	Blue mould rot	*Penicillium expansum* (Lk.) Thom
Banana	Crown rot	*Colletotrichum musae* (Berk. and Curt.) Arx (= *Gloeosporium musarum* Cke. and Mass.) *Fusarium roseum* Link amend. Snyd. and Hans. *Verticillium theobromae* (Turc.) Hughes *Ceratocystis paradoxa* (Dade) Moreau (= *Thielaviopsis paradoxa* [de Seynes] Höhn.)
	Anthracnose	*Colletotrichum musae* (Berk. and Curt.) Arx (= *Gloeosporium musarum* Cke. and Mass.)
Citrus fruit	Stem-end rot	*Phomopsis citri* Faw. *Diplodia natalensis* P. Evans *Alternaria citri* Ell. and Pierce
	Green mould rot	*Penicillium digitatum* Sacc.
	Blue mould rot	*Penicillium italicum* Wehmer
	Sour rot	*Geotrichum candidum* Lk ex Pers.
Grape, apple, pear, strawberry, leafy vegetables	Grey mould rot	*Botrytis Cinerea* Pers. ex. Fr.
Papaya, mango	Anthracnose	*Colletotrichum gloeosporiodes* (Penz.) Sacc.
Peach, cherry	Brown rot	*Monilinia fructicola* (Wint.) Honey (= *Sclerotinia fructicola* [Wint.] Rehm)
Peach, cherry, strawberry	Rhizopus rot	*Rhizopus stolonifer* Ehr. ex Fr.)
Pineapple	Black rot	*Ceratocystis paradoxa* (Dade) Moreau (= *Thielaviopsis paradoxa* [de Seynes] Höhn.)
Potato, leafy vegetables	Bacterial soft rot	*Erwinia carotovora* (Jones) Holland and other species
	Dry rot	*Fusarium* spp.
Sweet potato	Black rot	*Ceratocystis fimbriata* Ellis and Halst. [= *Endoconidiophora fimbriata* (Ell. and Halst.) Davidson]
Leafy vegetables, carrot	Watery soft rot	*Sclerotinia sclerotiorum* (Lib). de Bary

[1] Adapted from Eckert, J.W. Control of postharvest diseases. Siegel, M.R.; Sisler, H.D., eds. Antifungal compounds. Vol. 1. New York Marcel Dekker; 1977. 269–352. Eckert, J.W.; Brown, G.E. Postharvest citrus diseases and their control. Wardowski, W.F.; Nagy, S.C.; Grierson, W. eds. Fresh citrus fruits. Westport. CT: AVI; 1986: 315–60.

THE INFECTION PROCESS

Fruit and vegetables are rotted by organisms that either infect the produce while still immature and attached to the plant or during the harvesting and subsequent handling and marketing operations. The infection process, particularly postharvest, is greatly aided by mechanical injuries to the skin of the produce, such as fingernail scratches and abrasions, insect punctures and cut stems. Furthermore, the physiological condition of the produce, the temperature, and the formation of periderm (see later in this Chapter), significantly affect the infection process and the development of the infection. It is important to know the pattern of the infection process in order that suitable treatment strategies may be developed to control or eliminate the infection.

Preharvest infection

Preharvest infection of fruit and vegetables may occur through several avenues, for example, direct penetration of the skin, infection through natural openings of the produce, and infection through damage. Several types of pathogenic fungi are able to initiate an infection on the surface of floral parts and on sound, developing fruit. The infection is then arrested and remains quiescent until after harvest when the resistance of the host decreases and conditions become favourable for growth, for example, the fruit commences to ripen or tissue senesces. Such 'latent' infections are important in the postharvest wastage of many tropical and subtropical fruits, for example, anthracnose of mango and papaya, crown rot of banana and stem-end rots of citrus. For instance, spores of *Colletotrichum* germinate in moisture on the surface of the fruit, and within several hours of germination the end of the germ tube swells to form a structure known as an appressorium, which may or may not penetrate the skin before the infection is arrested.

Weak parasitic fungi and bacteria may also gain access to immature fruit and vegetables through natural openings, such as stomates (Figure 24), lenticels and growth cracks. Again these infections may not develop until the host becomes less resistant to the invading organism, for example, when the fruit ripens. It appears that sound fruit and vegetables can suppress the growth of these organisms for a considerable time, but little is known about the interaction of the invading microorganism and the host tissue. An example of this infection mechanism is the penetration of apple lenticels before harvest by spores of *Phlyctaena vagabunda*, which then manifest themselves in storage as rots round the lenticels.

Many pathogenic microorganisms present on dead plant tissue or associated with the soil can only infect produce through surface injuries, and often require favourable weather conditions at the time of maturation or ripening of the crop, or both, to cause major loss.

Postharvest infection

Many fungi that cause considerable wastage of produce are unable to penetrate intact skin of produce, but readily invade via any break in the skin. The damage is often microscopic but is sufficient for pathogens present on the crop and in the packing house to gain access to the produce. In addition, the cut stem is a frequent point of entry for microorganisms, and stem-end rots are important forms of postharvest wastage of many fruits and vegetables. Postharvest infection can also occur through direct penetration of the skin by, for instance, *Sclerotinia* and *Colletotrichum*.

Factors affecting development of infection

Probably the most important factor affecting the development of postharvest wastage of produce is the surrounding environment. High temperatures and high humidities favour the development of postharvest decay, and chilling injury generally predisposes tropical and subtropical produce to postharvest decay. In contrast, low temperature, low oxygen and high carbon dioxide levels and the correct humidity can restrict the rate of postharvest decay by either retarding the rate of ripening or senescence of produce, depressing growth of the pathogen, or both.

Many other factors affect the rate of development of an infection in fruit and vegetables. The host tissue, particularly the pH of the tissue, acts as a selective medium; fruits generally have a pH below 4.5 and are largely attacked and rotted by fungi; many vegetables have a pH above 4.5 and consequently bacterial rots are much more common. Ripening fruit is more susceptible to wastage than immature fruit, so treatments, such as low temperature, that slow down the rate of ripening will also retard growth of decay organisms. The underground storage organs, such as potato, cassava, yam, and sweet potato, are capable of forming layers of specialized cells (wound-periderm) at the site of injury, thus restricting the development of postharvest decay. During commercial handling of potato, periderm formation is promoted by ten to fourteen days storage at 7 to 15°C and 95 per cent relative humidity, a process known as curing. A type of curing process (possibly by desiccation) has been claimed to reduce the wastage of orange by *P. digitatum*: the fruit is held at high temperature (30°C) and humidity (90 per cent) for several days. The orange peel becomes less turgid under these conditions and lignin is synthesized in the injured flavedo tissue.

CONTROL OF POSTHARVEST WASTAGE

Preharvest

In most instances, control of postharvest wastage should commence before harvest in the field or orchard. Wherever possible, sources of infection should be eliminated, and sprays for the control or eradication of the causal organisms applied. Preharvest sprays are generally not as effective as postharvest application of the chemical directly to the commodity, although some of the systemic fungicides have shown good control of latent infections, such as lenticel rot of apple and brown rot of peach. With some of the newer, more specific fungicides, development of resistance to the fungicide by the organism has occured. For instance, the rapid development of resistance by *Penicillium* species to the benzimidazole group of fungicides strongly suggests that preharvest sprays of these fungicides would be unwise because of the opportunity for selection and growth of resistant strains, particularly if the same fungicide was being relied upon for postharvest control of *Penicillium*.

Careful handling during harvesting can minimize mechanical demage and so reduce subsequent wastage due to microbial attack. Similarly it is also unwise to harvest some fruits such as citrus after rain or heavy dew, as the peel is turgid and easily damaged.

Postharvest

Many physical and chemical treatments have been used for postharvest wastage control in fruit and vegetables. The effectiveness of a treatment depends on three main factors:

1. the ability of the treatment or agent to reach the pathogen;
2. the level and sensitivity of the infection; and
3. the sensitivity of the host produce.

The time of infection and the extent of development of the infection are critical in respect to whether it can be controlled. For example, *Penicillium* and *Rhizopus* invade wounds during harvest and subsequent handling operations and are much more easily controlled by fungicide application to the surface of the commodity than the grey mould of strawberry, which infects the fruit in the field some weeks before harvest or even at the time of flowering (Table 22).

Physical treatments

Postharvest wastage of produce may be controlled by low and high temperatures, modified atmospheres, correct humidity, magnetic fields, ionizing radiations, good sanitation and development of wound barriers. Low temperature handling and storage is the most important physical method of postharvest wastage control, and the remaining methods can be considered as supplements to low temperature. The extent to which low temperature and other environmental modifications can be used to control wastage is dependent on the tolerance of the tissue to that environment. For example, most tropical and subtropical produce is susceptible to chilling injury and therefore cannot be subjected to low temperature. Heat treatments, either moist hot air or hot water dips, have had limited commercial application for control of postharvest wastage in papaya, stone fruits and cantaloupe. The advantage of hot water dipping is that it can control surface infections as well as infections that have penetrated the skin and it leaves no chemical residues on the produce. The absence of chemical residues demands that recontamination of the produce by microorganisms be prevented by strict hygiene and possibly application of a fungicide, although at much lower levels than those required without a hot water dip. Hot water dips must be precisely administered as the range of temperature (50–55°C) necessary to control wastage approaches temperatures that damage produce. Ionizing radiations are effective in inhibiting microbial growth, but can cause physiological damage and aberrant ripening (Chapter 10).

Chemical treatments

Chemical control of postharvest wastage has become an integral part of the handling and successful marketing of fruit only during the past twenty-five years, particularly in the development of the world trade in citrus, banana and grapes. The level of control of wastage depends on the marketing strategy for the commodity and the type of infection. For citrus, which have a relatively long postharvest life, the aim of the treatment is to prevent primary infection and also sporulation so that nearby fruits are not contaminated. The strawberry has a short postharvest life, and treatment is aimed at preventing the spread of grey mould, which infected the strawberries in the field. In other words the treatment has to match the subsequent marketing of the commodity—there is no point in treating a short-life commodity with a fungicide that has a long residual activity. The success of a chemical treatment for wastage control depends on several factors:

1. the initial spore load;
2. the depth of the infection within the host tissues;
3. growth rate of the infection;
4. temperature and humidity; and
5. the depth to which the chemical can penetrate the host tissues.

Moreover, the applied chemical must not be phytotoxic, that is, injure the host tissues, and should fall within the ambit of the local food additive laws.

A wide range of chemicals has been, and still is, used for the control of postharvest wastage in fruit, particularly in citrus, banana, grapes and strawberry. Table 23 lists some of these compounds, their common names, the pathogens against which they are effective and the fruit on which they are applied. The chemicals listed in Table 23 are generally fungistatic in action rather than fungicidal; that is they inhibit spore germination or reduce the rate of germination and growth after germination rather than cause death of the organism and must come into direct contact with the organism to be effective. A few chemicals, such as chlorine and sulphur dioxide (SO_2), are true fungicides: chlorine is commonly added to wash-water to kill bacteria and fungi and sulphur dioxide is lethal to *Botrytis* on grapes. Chemicals may be applied impregnated into wraps or box liners, as fumigants, solutions and suspensions, or in wax.

Development of chemical treatments for control of citrus wastage

An instructive example of the development of postharvest chemical control is given for citrus. Green (*Penicillium digitatum*) and blue (*Penicillium italicum*) moulds are the major postharvest diseases of citrus, with green mould more prevalent in humid coastal areas; the stem-end rots (Table 22) are also significant causes of postharvest losses in more humid climates. Borax and sodium carbonate were the first chemicals to provide a measure of wastage control; both of these treatments were largely superseded by the more effective compound, sodium o-phenylphenate (SOPP), in the 1950s. Before 1940, Tomkins in England had found that the undissociated form of SOPP, o-phenylphenol (HOPP), con-

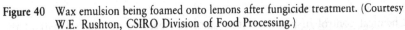

Figure 40   Wax emulsion being foamed onto lemons after fungicide treatment. (Courtesy W.E. Rushton, CSIRO Division of Food Processing.)

trolled citrus wastage, but it caused serious scalding of the skin. The advantage of SOPP was that it was not phytotoxic but was converted to the fungitoxic free phenol at the appropriate sites. The o-phenylphenate anion (OPP) diffuses selectively into injury sites and is converted to the undissociated form, preventing infection at these sites during storage or marketing. The SOPP–OPP system has broad spectrum activity and provides excellent control of *Penicillium* and the stem-end rots, but some resistance of *P. digitatum* to OPP has been reported where treated lemons have been stored for lengthy periods. A typical treatment for citrus consists of a two-minute dip in a 2 per cent solution of SOPP tetrahydrate at temperatures up to 32°C and pH of 11.7 or higher. SOPP may also be foamed (Figure 40) onto the fruit or incorporated into either wax applied to the fruit as further protection after washing (Figure 41) and/or an initial fungicide treatment (Figure 42).

The next major advance was the discovery of the fungistatic action of biphenyl against *Penicillium* and several other types of fruit wastage pathogens. Biphenyl has greatly assisted the development of world trade over the past forty years. Biphenyl, impregnated into fruit wraps or into paper sheets placed in the fruit container, sublimes into the atmosphere surrounding the fruit, preventing sporulation of *Penicillium* on the surface of infected fruits and thus the transfer of the infection to neighbouring fruits. Biphenyl wraps are often used as a complementary treatment on export fruit that have been treated with SOPP. Some problems are associated with biphenyl treatment: the fruit have a characteristic 'hydrocarbon' odour which tends to disappear within a few days after removal from the wraps; residues on the fruit surface sometimes exceed the permissible limit which varies between 70–100 micrograms per gram and strains of *Penicillium* and *Diplodia* have been able to develop resistance to biphenyl in areas

Figure 41 Washing lemons before application of fungicide to control citrus green mould. (Courtesy W.E. Rushton, CSIRO Division of Food Processing.)

Figure 42 Application of fungicide to citrus fruits by means of overhead irrigation. (Courtesy W.E. Rushton, CSIRO Division of Food Processing.)

Table 23 Chemicals that have been used as postharvest fungicides[1]

Name and formulation	Pathogen controlled	Host	Remarks
Alkaline inorganic salts			
sodium tetraborate (borax)	*Penicillium*	Citrus	Only reasonably effective; problem with B residues
sodium carbonate	*Penicillium*	Citrus	Only slightly effective
sodium hydroxide	*Penicillium*	Citrus	Only slightly effective, caustic
Ammonia and aliphatic amines			
ammonia gas	*Penicillium, Diplodia*	Citrus	Good for fumigation of degreening and storage rooms
	Rhizopus	Peach	Slight control
sec-butylamine	*Penicillium*, stem-end rots	Citrus	Good control as dip or fumigant
Aromatic amines			
dicloran	*Rhizopus, Botrytis*	Stone fruits, carrot, sweet potato	Very effective
Benzimidazoles			
benomyl, thiabendazole thiophanate methyl, carbendazim	*Penicillium*	Citrus	Effective at low concentration; resistance a problem; residue tolerance 0–10 μg/g
	Colletotrichum, other fungi	Banana, apple, pear, pineapple, stone fruits	
Triazoles			
imazalil	*Penicillium*, stem-end rots	Citrus	Effective against benzimidazole-resistant strains and at low concentration
prochloraz	*Penicillium*	Citrus	Effective against benzimidazole-resistant strains
Guanidine			
guazitine	*Penicillium*, Geotrichum	Citrus	Effective against benzimidazole-resistant strains
Hydrocarbons and derivatives			
biphenyl	*Penicillium, Diplodia*	Citrus	Smell unpleasant
methyl chloroform	*Penicillium*, stem-end rots	Citrus	Inhibits spore germination only

		Produce	
Oxidising substances			
hypochlorous acid	Bacteria, fungi build-up in wash water	Produce	Good sterilant, no penetration of injury sites, corrosive to metal
iodine	Bacteria, fungi	Citrus, grapes	Staining problem, expensive
nitrogen trichloride	*Penicillium*	Tomato, citrus	Hydrolyzes to hypochlorous acid
Organic acids and aldehydes			
dehydroacetic acid	*Botrytis* and other fungi	Strawberry	Dip not accepted by industry
sorbic acid	*Alternaria, Cladosporium*	Fig	
formaldehyde	Fungi		Sterilant for picking boxes, storage rooms
Phenols			
o-phenylphenol (HOPP)	*Penicillium*	Citrus	Causes fruit injury
sodium o-phenylphenate (SOPP)	*Penicillium*, bacteria, fungi	Produce	pH control needed to prevent injury; residue tolerance 10–12 µg/g
Salicylamilide	*Penicillium, Phomopsis, Nigrospora*	Citrus, banana	Slight control
Sulphur (inorganic)			
sulphur dust	*Monilinia Sclerotinia*	Peach	
lime-sulphur	*Botrytis*	Grapes	
sulphur dioxide gas, bisulphite		Grapes	Sulphur dioxide gas needs moisture to be effective; inexpensive; no toxic residues
Sulphur (organic)			
captan	Storage rots	Various produce	
thiram	*Cladosporium*, crown and stem-end rots	Strawberry, banana	
ziram	*Alternaria*, crown and stem-end rots	Banana	
thiourea	*Penicillium* spores	Citrus	Toxic to man
thioacetamide	*Diplodia*		

[1] Adapted from Eckert (1977).
Ogawa, J.M.; Manji, B.T.; El Behadli, A.H. Chemical control of postharvest diseases. Sharpley, J.M.; Kaplan, A.M. eds. Proceedings of the third international bio-degradation symposium, 1975, Kingston, RI: London; Applied Science; 1976; 561–75.
Wild B.L. Fungicides for postharvest wastage control in fruit marketing. Food Technol. Aust. 27: 477–8; 1975
Eckert, J.W.; Brown, G.E. Postharvest citrus diseases and their control. Wardowski, W.F.; Nagy, S.C.; Grierson, W. eds. Fresh citrus fruits. Westport, CT: AVI; 1986: 315–60.

where lemons were stored for up to four months. As a result biphenyl is banned in some countries.

Although SOPP gives excellent control of citrus wastage organisms, the conditions of application—maintenance of solution pH and strength, and rinsing with water after SOPP application must be strictly adhered to in order to avoid rind injury—presented some difficulty for packing houses. They welcomed the appearance of the benzimidazole group of fungicides twenty years ago. Thiabendazole (TBZ), benomyl, thiophanate methyl and carbendazim have a wide spectrum of anti-fungal activity at extremely low concentrations, but are inactive against *Rhizopus, Alternaria, Geotrichum* and soft rot bacteria. They are not phytotoxic and have low mammalian toxicity; in fact, TBZ was originally introduced as an anthelmintic (drench for worm control in stock) in 1961. They show systemic activity, that is, the compounds may be distributed round the plant by the transpiration stream. Thiabendazole is absorbed and transported as TBZ, however, benomyl and thiophanate methyl hydrolyze to carbendazim in water or in the plant and are translocated as carbendazim. It is believed that carbendazim provides most of the anti-fungal activity of these fungicides. The mode of action of these benzimidazole compounds appears to be similar, because they are effective against the same range of fungi. The benzimidazoles inhibit spore germination, interfere with mycelial growth and affect conidia formation even if the fruit are treated too late to control infection. It is thought that the benzimidazoles interfere with aspects of DNA (deoxyribonucleci acid) synthesis and cell replication, and this specificity of action appears to be responsible for the development of resistant strains of *Penicillium*.

The benzimidazoles are not readily soluble in water and are marketed as wettable powder formulations. Citrus fruits are generally flood irrigated with a suspension of between 0.05 and 0.1 per cent of the selected benzimidazole for 30 seconds to ensure thorough contact of the fungicide with sites of injury. The benzimidazoles have substantial residual activity—for example, benomyl is able to provide excellent wastage control in lemons stored for 6 months at 13°C. Because of the ease with which these compounds may be handled, packing houses rapidly adopted the benzimidazoles, but development of resistant strains of *Penicillium* has caused some packing houses to revert to SOPP. Therefore, the search for better fungicides has continued. Newer compounds which are effective against benzimidazole resistant fungi include *sec*–butylamine (2–aminobutane, 2–AB), the triazoles (imazalil, prochloraz) and guanidine (guazitine). *Sec*–butylamine is effective against most citrus wastage organisms and has considerable residual activity. Fruit may be fumigated or flood irrigated with *sec*–butylamine. Imazalil, prochloraz and guazitine are water soluble.

Development of chemical treatments to control wastage in other fruits

The storage and transportation of other fruits benefit from suitable selection and application of chemical fungicides. For example, 'black-end', 'finger-stalk' or 'crown' rot of banana and brown rot in peach are effectively controlled by the benzimidazoles; grey mould in table grapes by the slow evolution of sulphur dioxide by an in-package generator or by room fumigation with sulphur dioxide; and Rhizopus rot in stone fruits by dicloran. In contrast, control of *Geotrichum, Alternaria* and the soft rots is still unsatisfactory, although SOPP does give some control of *Geotrichum* when applied 4 to 24 hours after infection. Furthermore, much more needs to be learnt about the mechanisms which lead to the appearance of strains resistant to some of the more potent, but quite specific in action, fungicides.

In summary, the ideal postharvest fungicide should: be water soluble; have broad spectrum activity; not be phytotoxic; be safe to use, that is, leave no residues toxic to consumers; not affect palatability; retain activity over a long period; leave no visible residues; and be cheap, that is, the compound should be inexpensive or effective at low concentration. None of the present fungicides meet all of these requirements.

FURTHER READING

Coursey, D.G.; Booth, R.H. The postharvest phytopathology of perishable tropical produce. Rev. Plant Pathol. 51: 751–65; 1972.

Dekker, J.; Georgiopoulos, S.G., eds. Fungicide resistance in crop protection. Wageningen: Pudoc; 1982.

Eckert, J.W.; Brown, G.E. Postharvest citrus diseases and their control. Wardowski, W.F.; Nagy, S.C.; Grierson, W., eds. Fresh citrus fruits. Westport, CT: AVI; 1986: 315–60.

Eckert, J.W.; Rubio, P.P.; Mattoo, A.K.; Thompson, A.K. Postharvest pathology. Part 2. Diseases of tropical crops and their control. Pantastico, E.B., ed. Postharvest physiology, handling and utilization of tropical and subtropical fruits and vegetables. Westport, CT: AVI: 1975: 415–43.

Erwin, D.C. Systemic fungicides: disease control, translocation, and mode of action. Annu. Rev. Phytopath. 11: 389–422; 1973.

Hall, E.G.; Scott, K.J. Storage and market diseases of fruit. Melbourne, Australia: CSIRO; 1977.

Harvey, J.M. Reduction of losses in fresh market fruits and vegetables. Annu. Rev. Phytopathol. 16: 321–41; 1978.

Kader, A.A. et al. Postharvest technology of horticultural crops, Berkeley, CA: University of California; 1985. Special Publication 3311.

Lieberman, M. Post-harvest physiology and crop preservation. New York: Plenum; 1983.

Meredith, D.S. Transport and storage diseases of bananas: biology and control. Trop. Agric. 48: 35–50; 1971.

Reuther, W.; Calavan, E.C.; Carman, G.E., eds. Citrus Industry. Vol IV. Diseases and injuries, viruses; registration, certification, indexing; regulatory measures, vertebrate pests; biological control of insects; nematodes. Berkeley, CA: University of California; 1978.

Smith, W.L. Non-chemical control of postharvest deterioration of fresh produce. Sharpley, J.M.; Kaplan, A.M., eds. Proceedings of the third international biodegradation symposium: 1975: Kingston, RI. London: Applied Science; 1976: 577–87.

Smoot, J.J.; Houck, L.G.; Johnson, H.B. Market diseases of citrus and other subtropical fruits. Washington, DC: US Department of Agriculture; 1971. Agriculture Handbook No. 398.

10
Commodity treatments

Efficient production, distribution and marketing of fruit and vegetables relies on extensive preharvest application of chemicals to control pests, diseases and weeds. The previous chapter described the widespread postharvest application of a range of chemicals to restrict microbial wastage of produce. Other forms of deterioration, such as sprouting, water loss, storage disorders, and insect infestation, can be minimized with chemical treatment, but to a lesser degree than microbial wastage. The most extensive postharvest chemical treatment, other than wastage control, is the use of ethylene, a natural plant hormone, for the ripening of fruit, particularly banana, and the degreening of citrus.

CONTROLLED RIPENING

Climacteric fruits, particularly tropical and subtropical species, are frequently harvested when less than fully ripe and then transported, often over considerable distances, to areas of consumption. Here these fruits are ripened to optimum quality under controlled conditions of temperature, relative humidity, and with some fruits, through the addition of ripening gases. A further advantage of controlled ripening is to improve uniformity of ripening of fruit. The use of relatively high ripening temperatures may also minimize the development of rots in ripe tropical fruits. In contrast, non-climacteric fruits generally undergo little or no desirable change in composition after harvest and must not be harvested until fit for consumption.

A significant proportion of the world production of bananas of approximately 18 million tonnes is ripened under controlled conditions. The banana is unusual in that it can be picked over a wide range of physiological ages from half-grown (thin) to fully grown, and ripened to excellent quality with the aid of ethylene. It has been long known that banana can be induced to ripen by enclosing the fruit in a chamber of some kind with restricted ventilation. In more recent times it was found that enclosing smouldering wood or charcoal with banana hastened ripening. It is now known that ethylene is a product of incomplete combustion of such fuels and that it is by far the most active of the known ripening gases. Acetylene, generated by adding water to calcium carbide, produces a ripening response, but in practice at least a concentration 100-times higher is required.

The commercial ripening of bananas has been reduced to a routine operation, and fruit at a specified colour stage (Table 24) can be produced on a predetermined schedule. The effective concentration of ethylene for banana ripening is quite low (Table 25), and when this concentration is maintained for the stated

Table 24   Colour stages of the ripening Cavendish banana[1]

Stage	Peel colour	Approx. Starch (%)	Approx. Sugar (%)	
1	Green	20	0.5	Hard, rigid; no ripening
Sprung	Green	19.5	1.0	Bends slightly, ripening started
2	Green, trace of yellow	18	2.5	
3	More green than yellow	16	4.5	
4	More yellow than green	13	7.5	
5	Yellow, green tip	7	13.5	
6	Full yellow	2.5	18	Peels readily; firm ripe
7	Yellow, lightly flecked with brown	1.5	19	Fully ripe; aromatic
8	Yellow with increasing brown areas	1.0	19	Over-ripe; pulp very soft and darkening, highly aromatic

[1] From Commonwealth Scientific and Industrial Research Organization. Banana ripening guide. Melbourne. Australia: 1972. Division of Food Research Circular 8.

Table 25   Ripening conditions for some fruits using ethylene[1]

Fruit	Temperature (°C)	Ethylene Concentration (μL/L)	Treatment Time (hours)
Avocado	18–21	10	24–72
Banana	15–21	10	24
Cantaloupe	18–21	Nil	
Honey Dew melon	18–21	10	24
Kiwi fruit	18–21	10	24
Mango	29–31	10	24
Papaya	21–27	Nil	
Pear	15–18	10	24
Persimmon	18–21	10	24
Tomato	13–22	10	continuously

[1] Relative humidity is normally maintained at 85 to 90 per cent.

Figure 43   Banana ripening rooms. (Courtesy W.E. Rushton, CSIRO Division of Food
Processing.)

period, further increases in concentration give no added advantage. In practice,
however, high concentrations are maintained because the ripening rooms are not
sufficiently airtight. Bananas may be ripened by a batch process in which the
chamber (Figure 43) is charged with ethylene gas to a concentration of between
20 and 200 micro-litres per litre. The chamber has to be ventilated after the first
24 hours to prevent the carbon dioxide concentration from exceeding 1 per cent
and thus retarding ripening. If the chamber is poorly sealed, it may be necessary
to recharge with ethylene after 12 hours. Fruit are removed at the desired colour
stage. A more satisfactory alternative to charging the ripening chamber with an
initial high concentration of ethylene is to trickle ethylene into the chamber at a
rate just sufficient to maintain the concentration shown in Table 25. The ripening chambers should be ventilated at the rate of about one room volume each six
hours to prevent the accumulation of carbon dioxide. In practice it is usually
not necessary to install a ventilation system in rooms less than 60 cubic metres
because they have natural air exchange (leakage) rates higher than the required
minimum rate. For larger rooms, the natural air leakage rates should be taken
into account when calculating the ventilation requirement. Excess ventilation
will increase energy consumption for cooling and heating.

   Traditionally, containers of bananas were stacked in a ripening room so as to
expose at least two faces to the circulating air to ensure that fruit temperatures

were even. In modern practice the fruit are packed in vented cartons, unitised on pallets, and fruit temperatures are controlled by forced-air circulation. Conversion of existing rooms to the forced-air system has increased the capacity of the rooms by about 40 per cent and enabled fruit temperatures to be controlled more accurately. A minimum air-flow of about 0.34 litres per second per kilogram of bananas is required. In Australia, the forced-air system is also commonly used in tomato ripening rooms.

Water loss can be high at ripening temperatures unless relative humidity is maintained at a high level. Humidity can be raised by atomizing water into the ripening chamber or simply by spraying the floor with water. Regulation of relative humidity during the course of ripening can be particularly important for banana. A relative humidity of 85 to 90 per cent has been recommended for ripening to stage 2 (Table 24) but this should be reduced to between 70 and 75 per cent during the later colouring stages. Although the best skin colour may be achieved at the highest relative humidity commercial experience with conventional chambers not designed for forced-air circulation has shown that the skin tends to be too soft and may split. If relative humidity is too low, weight loss may be excessive, colour poorer and blemishes more pronounced. Maintenance of relative humidity at 85 to 90 per cent continuously in forced-air ripening rooms has proved satisfactory. Because of the high relative humidity and temperature maintained in ripening rooms, moulds grow readily on any organic matter, including the walls of the room if they are not suitably protected. Regular cleaning with a solution of sodium hypochlorite (chlorine) followed by fumigation with formaldehyde gas increases the life of ripening rooms, reduces maintenance costs, and minimizes fruit wastage.

The ripening of other climacteric fruits that have been harvested immature can be hastened by treatment with ethylene under controlled conditions, but in contrast to banana and avocado, quality will be inferior to that of fruits harvested at the mature green stage. With many of these fruits, it is more important to harvest at the correct stage of maturity—for example the full slip stage for cantaloupe, the first appearance or yellow colour in the blossom end of papaya, and the first colour 'breaker' stage of tomato—and then it is only necessary to hold at the temperature and relative humidity specified in Table 25 to achieve a high quality ripened fruit. In other words, treatment with ethylene is unnecessary for these fruits to produce fully ripened produce provided they are picked when sufficiently developed, although ethylene treatment will promote more uniform ripening in consignments of mixed maturities.

CONTROLLED DEGREENING

The pulp of many early-season citrus cultivars becomes edible before the green colour has completely disappeared from the peel. Exposure to low temperature during maturation is necessary for the development of an orange-coloured peel; this requirement explains why the peel of citrus grown in the low altitude tropics fails to degreen completely. Furthermore, the Valencia orange cultivar is often stored on the tree for several months after ripening has been completed, and during this storage period the peel tends to regreen. Treatment with ethylene under controlled conditions hastens the loss of chlorophyll: this process is known as degreening. Batch or trickle degreening is a 'cosmetic' treatment designed to give the fruit a ripe appearance, but does not result in significant changes in pulp composition if correctly administered. The conditions of batch degreening—20 to 200 microlitres per litre ethylene, 25 to 30°C and 90 to 95 per cent relative humidity—are maintained for two to three days with regular ventilation of the chamber to prevent buildup of carbon dioxide (citrus is injured by carbon dioxide concentrations above 1 per cent). Trickle degreening, with 10 microlitres per litre ethylene continuously metered into the room, is more rapid than batch degreening and is, therefore, preferable as the conditions of degreening accelerate deterioration and wastage of citrus. Although the most rapid degreening occurs at 25 to 30°C, production of peel carotenoids is greatest at 15 to 25°C.

Recent research has shown that fruit may be ripened and degreened equally as well with the ethylene-releasing compound, ethephon, as with ethylene gas. Ethephon is absorbed by fruit tissues and when the pH exceeds 4.6, breaks down to release ethylene.

In some citrus growing areas, notably Florida in the USA, tangelos, temples and some early season oranges, have not experienced enough cold weather to promote the development of a highly coloured peel. Packing sheds in these areas are legally permitted to dye the peel of these fruits under strictly controlled conditions with Citrus Red number 2: the process is known as 'colour-add' and can only be applied to mature fruits that are not intended for processing.

CONTROL OF SUPERFICIAL SCALD IN APPLE

This important disorder of apple in cool storage is now known to be caused by the oxidation products of α-farnesene that occur when the natural antioxidants in the fruit are degraded or inactivated during cool storage (Chapter 7). The addition of various synthetic antioxidants effectively prevents the oxidation of α-farnesene and hence the development of scald. Commercially, scald has been significantly reduced by a postharvest dip in either diphenylamine (0.1–0.25 per cent) or 1.2–dihydro–6–ethoxy–2.24–trimethylquinoline (ethoxyquin) (0.2–0.5 per cent), diphenylamine may also be applied by means of impregnated wraps or in wax, or both antioxidants may be brushed onto the fruit. Both compounds were discovered within a few years of each other, and most countries have approved one or the other, but rarely both. This has caused problems in international trade in apples, where exporting countries have to use the scald treatment that is legally permitted by the importing nation. As the treatment is applied immediately after harvest, the final destination of the fruit must be known with some certainty at harvest, and some segregation of treated fruit must be maintained during storage.

Although diphenylamine and ethoxyquin adequately control scald, a single compound, approved by all countries, is desirable. For political reasons, it is unlikely that either compound will gain universal approval. Chemical companies are reluctant to pursue approval for various other compounds that have been shown to be effective scald inhibitors due to increasing demands of health authorities. The dilemma over scald treatments highlights the problems that can be expected more often in the future. There is a reluctance internationally to allow new chemicals as food additives, and this particularly affects the postharvest treatment of produce as food laws are mainly concerned with foods after harvest. Fewer restrictions are placed on preharvest treatments as a wide range of toxic pesticides are allowed on most crops, although strict residue limits are now being established for many of these agricultural chemicals.

CALCIUM APPLICATION TO APPLE

Since about 1960, preharvest calcium sprays have been applied commercially to apple in several countries to reduce the incidence of bitter pit after harvest. The added calcium also reduces internal breakdown of apple in cool storage. To be effective, calcium must actually contact the fruit and be absorbed directly by it; calcium that falls on the leaves or branches is not transferred to the fruit. The sprays may be applied up to six times during the growing season to slowly build-up the calcium level in the fruit, as it is difficult to ensure that all fruit are wetted sufficiently by one application of calcium solution. This problem may be overcome by dipping apples after harvest in solutions of calcium salts. A further improvement has been developed whereby a partial vacuum or positive pressure is applied to the solution, which force calcium solution into the apple flesh and thus control bitter pit and breakdown much better than preharvest sprays or the postharvest dip alone.

New Zealand has been using the vacuum infiltration technique on a commercial scale for several years with considerable success. Its use in other countries has shown less promise. It seems that best results are obtained with apples that have a closed calyx so that the calcium solution is forced into the fruit through the lenticels and is thus spread around the perimeter tissue where the disorders occur. With open calyx fruit, the uptake of solution is difficult to control as it readily enters the fruit via the calyx and excess solution accumulates in the core area, often leading to injury or rotting.

Laboratory studies with vacuum infiltration of calcium solutions have shown that the technique can markedly retard the initiation of ripening in a number of climacteric fruit such as tomato, avocado, mango and pear. However, larger scale experiments have not yet been reported for any commodity so that the commercial potential of the treatment is not known. In addition to showing that the treatment gives a consistent quantitative delay in ripening, it is necessary to adequately examine the potential problems of skin injury due to excess calcium uptake and rot development.

INCREASING WATER LOSS FROM APPLE

The incidence of low temperature breakdown in apple is reduced by increasing the rate of water loss from the fruits. This is most easily achieved by storing loose fruits in air of low relative humidity; or, if fruits are packaged, by creating a low relative humidity atmosphere within each package by including a water absorbent with the fruits. The latter technique could be more desirable where mixed varieties are stored in the same room, as the varieties that are not susceptible to breakdown would still require air of high relative humidity to minimize transpiration.

Anhydrous sodium carbonate is an ideal absorbent as it is non-hygroscopic but absorbs water uniformly from fruit and can be also added to fruit wraps. No commercial operator has deliberately exploited the technique of increased water loss although some cool store operators would have unintentionally applied the low relative humidity treatment to fruit in poorly controlled rooms.

WAXING

Commodities that have a waxy skin lose water slowly, and this observation led to the application of wax to certain commodities, particularly citrus fruits, which shrivel rapidly and lose consumer appeal during storage and marketing. Most waxes currently in commercial use are proprietary formulations, which may contain a mixture of waxes that are derived from plant and petroleum sources. Many of these formulations have been based on a combination of paraffin wax, which gives good control of water loss but a poor lustre to the produce, and carnauba wax, which imparts an attractive lustre to the produce but provides poorer control of water loss. In recent years formulations containing polyethylene, synthetic resin materials, emulsifying and wetting agents have become popular. The wax formulation is often used as a carrier of fungicides, and inhibitors of senescence, superficial scald and sprouting.

Waxes are applied to produce for the two-fold purpose of reducing water loss and, therefore, the rate of shrivelling, and of improving the appearance to the consumer. The rate of water loss can be reduced by 30 to 50 per cent under commercial conditions, particularly if the stem scar and other injuries are coated with wax. Waxes are brushed, sprayed, fogged, or foamed (Figure 41) onto produce, or produce is conveyed through a tank of wax emulsion. The wax film must be thin, or gas exchange may be hindered. Following application of wax, produce is generally dried and polished.

As mentioned, most citrus is waxed, because washing can remove much of the natural wax from the peel thus exacerbating shrivelling and loss of appearance. Many other commodities, such as cucumber, tomato, passionfruit, pepper, banana, apple and some root crops, are now being waxed to reduce weight loss and to increase their sales appeal.

PLANT GROWTH REGULATORS

The five groups of plant growth regulators—auxins, gibberellins, cytokinins, abscisic acid and ethylene—control many plant functions and biosynthetic pathways and have been ascribed a regulatory role in all aspects of fruit and vegetable development from germination or primordia formation to senescence. Information on the effect of the first four groups of compounds on mature fruit and vegetables is not as extensive as that for ethylene, but they are worthy of further study as they are effective at extremely low concentrations, and since many occur naturally, their acceptance as potential food additives should be easier.

Few commercial postharvest applications of these plant growth regulators, except ethylene, have been implemented. Two examples are the retardation of button senescence on citrus fruit by 2,4–dichlorophenoxyacetic acid (2,4–D) to prevent the development of the stem-end rots, and the retardation of the senescence of the peel of citrus fruit by gibberellic acid (GA_3). The cytokinins, notably the synthetic cytokinin benzyladenine, significantly delay the loss of chlorophyll from, and the general senescence of, many green leafy vegetables. However, application for their use on leafy vegetables in the USA was not approved.

SPROUT INHIBITORS

The buds of potato and onion, at maturity, enter a dormant state through which the plant can survive periods unfavourable to plant growth; they are normally harvested at this time and can be stored for many months under correct conditions for either retail marketing or processing.

In potato the duration of postharvest dormancy—commonly known as the rest period—is influenced by preharvest factors, maturity and variety, but generally not the temperature of storage. Once the rest period ends, the rate of sprouting is dependent on temperature. Sprouting of potato rarely occurs below 4°C, but storage at these temperatures is impractical due to the conversion of starch to sugars (Chapter 4). At temperatures greater than 4°C sprouting is a problem during storage periods greater than two to three months.

The application of several chemicals and ionizing radiation (see later this Chapter) effectively suppress sprouting in potato and onion during storage at higher temperatures. Maleic hydrazide (MH), nonyl alcohol, 3–chloroisopropyl–N–phenylcarbamate (CIPC), isopropyl–phenylcarbamate (IPPC), methyl naphthaleneacetic acid (MENA) and 2,3,4,6–tetrachloronitrobenzene (TCNB) are commercial sprout inhibitors, but legal restrictions on the usage and permitted residues of these compounds vary among countries. CIPC is a strong sprout inhibitor and is probably the most widely used of the chemical sprout inhibitors for the storage of potato. CIPC may be applied as a dust, water dip, vapour, or aerosol and, like many of the other chemical sprout inhibitors, interferes with periderm formation and thus should be applied after curing (Chapter 9). In the USA the tolerance for CIPC in raw and processed potato is 50 micrograms per gram. Sprouting of onion during long term storage is effectively prevented by MH applied several weeks before harvest.

DISINFESTATION

Some insect species, particularly the tephritid fruit flies, that attack horticultural crops can seriously disrupt produce trade between countries. Table 26 lists the more important species, their distribution, and hosts. The importing country often erects a quarantine barrier against produce originating from an area in which the insect species of concern is known to occur: for example, the distribution of the Mediterranean fruit fly is shown in Figure 44. In order to market produce, the exporting nation must develop a disinfestation treatment that satisfies the importing country with respect to the survival of the eggs, larvae, and pupae of the insect, but one that will not harm the produce or the consumer. It must also be economical to apply. The occurrence of these insects may also restrict movement of produce within a country.

Several disinfestation procedures, employing chemical, physical, or irradiation treatments, have been developed to eliminate insects from produce. Of these treatments, three are in commercial use and are acceptable to the quarantine authorities of various importing nations: they are fumigation with gaseous sterilants, storage at low temperature, and a short exposure to high temperature. Very few governments have approved ionizing radiation as a disinfestation treatment for fruit.

Fumigation with gaseous sterilants is the most important technique for disinfesting produce. Many gaseous sterilants, such as acrylonitrile, carbon disulphide, carbon tetrachloride, ethylene dibromide (EDB), ethylene oxide, hydrogen cyanide, methyl bromide, phosphine, and sulphuryl fluoride, are available and used for general quarantine fumigations, but methyl bromide is the most widely used fumigant for produce. Effective fumigation was provided in the past by EDB, carbon disulphide and hydrogen cyanide. Carbon disulphide and hydrogen cyanide fell into disfavour because of their flammability and high toxicity to humans, respectively. Use of EDB as a fumigant on produce for consumption in the USA was prohibited from 1 September 1984 because of its suspected carcinogenicity. Some other countries have since banned its use.

Methyl bromide may be applied to produce either in a permanent fumigation chamber or in a temporary enclosure, such as under a tarpaulin, and in gas-tight rail cars and road trucks. The permanent chamber offers the safest and most reliable fumigation; fumigation under tarpaulins offers flexibility at the dockside or railway yard, and if correctly carried out is as effective as chamber fumigation. Complete operating details for disinfesting produce by chemical and physical procedures are found in the *Plant Protection and Quarantine Treatment Manual* prepared by the Animal and Plant Health Inspection Service of the US Department of Agriculture; these or similar procedures should be carefully followed to give reliable treatment, avoid injury to the produce, and protect the operators.

Minimum effective doses of chemical disinfestants or treatments by physical methods that leave no survivors have been established for a range of insects and horticultural commodities (Figure 45). Maximum permissible residues of chemical disinfectants are specified by law in most countries.

Some fruits and vegetables are quite sensitive to the chemical disinfestants and alternative disinfestation procedures have been developed, mainly using high and low temperatures. During the last few years, the commercial use of these treatments has increased because authorities and consumers have become more concerned about chemical residues in produce. Many insects that infest

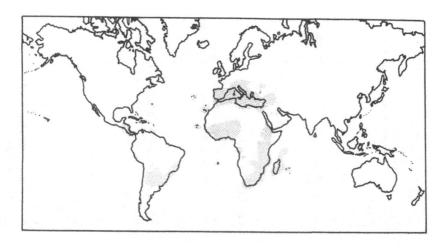

Figure 44   Distribution of Mediterranean fruit-fly. (*Ceratitis capitata* Wied.) (*redrawn from Distribution Maps of Pests*, Map No. 1, Commonwealth Institute of Entomology, London.)

produce do not tolerate exposure to low temperatures, an observation that has led to an effective disinfestation procedure for deciduous fruits but not for tropical and subtropical fruits, which are liable to chilling injury. For instance, US Quarantine authorities stipulate the following cold treatments for produce from areas infested with the Mediterranean fruit fly:

ten days at 0°C or below
eleven days at 0.6°C or below
twelve days at 1.1°C or below
thirteen days at 1.7°C or below
sixteen days at 2.2°C or below

To guarantee disinfestation against both Queensland and Mediterranean fruit flies, the Japanese Quarantine authorities have accepted cold treatment of Australian oranges for 16 days below 1.0°C, measured with a tolerance of 0.6°C. Similar treatments are being developed for other cold tolerant fruits with potential for export to Japan and other countries. Some importing countries permit cold treatments to be applied during transit. Cold treatments may also be applied in combination with chemical fumigation, thus reducing the amount of fumigant required.

Produce can also be successfully disinfested by exposure to high temperature —a process known as vapour-heat treatment. The temperature of citrus fruits, mango, papaya, pineapple, and some vegetables is raised to 43°C with air saturated with water vapour and maintained at 43°C for six to eight hours.

Table 26   Some insects and mites that can be carried by fruit and vegetables

Insect	Common name	Common hosts[1]	Approximate distribution[1]
Cryptophlebia leucotreta Meyr.	False codling moth	Citrus, avocado, stone fruit, guava	Africa
Cydia pomonella (L.)	Codling moth	Apple, pear, peach, quince, *Prunus* species, walnut	World-wide
Cylas formicarius (Fab.)	Sweet potato weevil	Sweet potato	Africa, Asia, Pacific Islands, North America, South America
Fruit flies, such as *Anastrepha fraterculus* (Wied.)	South American fruit fly	Peach, guava, citrus, *Spondias* species, *Eugenia* species	South America, West Indies, Central America
A. ludens (Lw.)	Mexican fruit fly	Citrus, other tropical and subtropical fruits	Central America, Mexico
Ceratitis capitata (Wied.)	Mediterranean fruit fly	Deciduous and subtropical fruits, especially peach and citrus	Southern Europe, Africa, Central America, South America, Western Australia, Hawaii
C. rosa Karsch	Natal fruit fly	Many deciduous and subtropical fruits	Africa
Dacus ciliatus (Lw.)	Lesser pumpkin fly	Cucurbits	Africa, India, Pakistan, Bangladesh
D. cucurbitae Coq.	Melon fly	Cucurbits, tomato	Asia, Hawaii, Papua New Guinea, Africa
D. dorsalis (Hend.)	Oriental fruit fly	Most fleshy fruits or vegetables	Asia, Hawaii
D. tryoni (Frogg.)	Queensland fruit fly	Many deciduous and subtropical fruits	Australia, Pacific Islands
Rhagoletis cerasi (L.)	Cherry fruit fly	Cherry, *Lonicera species*	Europe
R. cingulata (Lw.)	Cherry fruit fly	Wild and cultivated cherry, *Prunus* species	North America
R. pomonella (Walsh)	Apple maggot	Apple, blueberry	USA, Canada
Graphognathus leucoloma (Boh.)	White fringed weevil	Root vegetables	South Africa, Australia, New Zealand, USA, South America

Table 26   (continued)

Insect	Common name	Common hosts[1]	Approximate distribution[1]
Halotydeus destructor (Tucker)	Red-legged earth mite	Leafy vegetables	Australia, New Zealand, Africa
Lobesia botrana (Schiff.)	Vine moth	Grape	Europe, Japan, Africa
Maruca testulalis (Geyer)	Bean pod borer, mung moth	Legumes	Africa, Asia, Australia, Central America, South America, Pacific Islands
Mealybugs, such as			
Planococcus citri (Risso)	Citrus mealy bug	Citrus, grape	World-wide
Dysmicoccus brevipes (Ckll.)	Pineapple mealy bug	Pineapple	Africa, Asia, Australia, Pacific Islands, South America
Panonychus ulmi (Koch)	European red mite	Apple and other deciduous fruits	Europe, Africa, Asia, Australia, New Zealand, North America, South America
Phthorimaea operculella (Zell.)	Potato tuber moth	Potato, tomato, egg-plant	World-wide
Scale insects, such as			
Aonidiella aurantii (Maskell)	Red scale	Citrus	World-wide
Lepidosaphes beckii (Newm.)	Purple scale	Citrus	World-wide
Quadraspidiotus perniciosus (Comst.)	San José scale	Deciduous fruits	World-wide
Sternochaetus mangiferae (Fab.)	Mango seed weevil	Mango	Africa, Asia. Australia

[1] Prepared from *Distribution maps of insect pests* issued by the Commonwealth Institute of Entomology, London.

IRRADIATION

The preservation of food by ionizing radiations has received considerable attention during the last forty years. Initially, some optimistic claims were made for the preserving properties of ionizing radiation, but these were based largely on improperly conducted trials. Research has subsequently refuted many of these earlier claims.

The potential of ionizing radiation in the postharvest handling of fruit and vegetables lies in two directions: the disinfestation of produce of fruit flies and other insect pests, and the extension of the shelf life of produce by restricting the growth of wastage organisms and the retardation of aspects of produce metabolism, such as ripening and sprouting. Before this potential can be realized, several criteria must be met:

1. The host must have a considerably higher tolerance to radiation than the organisms or metabolic systems causing deterioration.
2. The required radiation treatment must be as, or more, economical than competing effective treatments.

Figure 45   Establishment of effective treatments for the disinfestation of produce. *Below:* Breeding fruit flies. *Above right:* Fruit flies laying their eggs in oranges. *Below right:* Apple being experimentally infested with light brown apple moth larvae. (Courtesy W.E. Rushton, CSIRO Division of Food Processing.)

ORANGES BEING INFESTED
BY FRUIT FLIES FOR EXPERIMENT

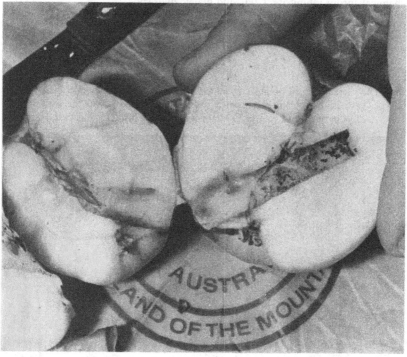

3. The radiation treatment must be acceptable to health authorities, that is, the wholesomeness of the irradiated produce must be demonstrated.

The disinfestation of fruit, particularly tropical fruits, such as papaya and mango, by ionizing radiation has been shown to be technically feasible, but less economic, than other procedures. The egg phase of the life cycle of insects is the most sensitive to irradiation, followed by larval, pupal and adult stages. Most insects are sterilized at doses of 0.05 to 0.2 kilogray; some adult moths will survive 1 kilogray, but their progeny are sterile. Despite extensive studies of the radiation disinfestation of papaya in Hawaii, very few governments have approved ionizing radiation as a disinfestation treatment for fruit, and no fruit is yet being so treated commercially. However, this may change with the introduction of restrictions on disinfestation by chemicals.

By 1985, thirty countries had issued provisional or conditional clearances covering the irradiation of over fifty different food items: twenty of these clearances were for sprout inhibition of potato and onion. At present, Japan makes the widest use of irradiation of potato to suppress sprouting (Figure 46). Sprouting in potato and onion during long term storage is inhibited by 0.02 to 0.15 kilogray, a dose which has little effect on other aspects of potato and onion quality, such as sugar level, rate of wastage and water loss, texture, and flavour. Irradiation of potato and onion is more expensive than treatment with the chemical sprout inhibitors, CIPC and maleic hydrazide.

Disinfestation and sprout inhibition are good examples of instances where the

Figure 46   Irradiation of potatoes to inhibit sprouting. Trays of potatoes pass through a field of irradiation from a source concentrated within the vertical rods. (Courtesy of Dr K. Vas, Central Food Research Institute, Budapest, Hungary.)

Table 27 Comparison of maximum tolerable and minimum radiation dose required for desired technical effects on selected fresh produce.[1]

Produce	Desired technical effect	Max. tolerable dose (kGY)[2]	Min. dose required (kGY)	Phenomena limiting commercial application
Apple	Control of scald and brown core	1–1.5	1.5	Cheaper, more effective alternatives; tissue softening
Apricot, peach, nectarine	Inhibition of brown rot	0.5–1	2	Tissue softening
Asparagus	Inhibition of growth	0.15	0.05–0.1	Economics, short season, small acreage
Avocado	Inhibition of ripening and rot	0.25	none	Cheaper, more effective alternatives; browning and softening of tissues
Banana	Inhibition of ripening	0.5	0.30–0.35	Cheaper, more effective alternatives
Lemon	Inhibition of penicillium rots	0.25	1.5–2	Severe injury to fruit at doses of 0.5 kGY or more; cheaper, more effective alternatives
Mushroom	Inhibition of stem growth and cap opening	1	2	Cheaper, more effective alternatives; no technical effect under commercial conditions
Orange	Inhibition of penicillium rots	2	2	Cheaper, more effective alternatives; no technical effect under commercial conditions
Papaya	Disinfestation of fruit fly	0.75–1	0.25	Not economical, inadequate acreage
Pear	Inhibition of ripening	1	2.5	Abnormal ripening; cheaper, more effective alternatives
Potato	Inhibition of sprouting	0.2	0.08–0.15	Cheaper, more effective alternatives
Strawberry	Inhibition of grey mould	2	2	Cheaper, equally effective alternatives
Table grape	Inhibition of grey mould	0.25–0.50		Tissue softening, severe off-flavours; cheaper, more effective alternatives
Tomato	Inhibition of alternaria rot	1–1.5	3	Abnormal ripening, tissue softening

[1] Adapted from Maxie, E.C.; Sommer, N.F.; Mitchell, F.G. Infeasibility of Irradiating fresh fruits and vegetables. HortScience 6: 202–4; 1971.
[2] 1 Gray = 100 rads.

difference in sensitivity to the low doses of ionizing radiation between the produce item and the organism or metabolic system to be restricted is sufficiently great for the irradiation process to be technically feasible. When produce, particularly fruit, is irradiated with intermediate levels of ionizing radiation of 0.5 to 2 kilograys to control the growth of common wastage organisms, the tissues of the fruit are often severely damaged. The most common effect of ionizing radiation is to soften the produce making it more susceptible to damage during subsequent handling and transport.

Maxie and his colleagues in California, after extensive trials with irradiated produce, were sceptical about the prospects of irradiation as a postharvest treatment as, even for those commodities that respond favourably, cheaper and often more effective alternative procedures are available (Table 27). Others claim that irradiation will become more attractive for certain items, such as potato and onion and high-cost tropical fruits for export, if the community objects to the continued use of chemicals for disinfestation, wastage control and sprout inhibition, and if the cost of irradiation can be lowered. Three joint FAO/WHO reports on the wholesomeness of irradiated foods have largely alleviated early fears about induced radioactivity, loss of nutritive value, or production of harmful radiolysis products in irradiated produce. The first of these reports (WHO, 1977) recommended acceptance of nine irradiated foods including potato, onion, mushroom, papaya, and strawberry. The second report (WHO, 1981) extended the list of approvals to include, amongst others, mango, and concluded that irradiation of any food commodity up to an overall average dose of 10 kGy presents no toxicological hazard and hence toxicological testing of foods so treated was no longer required. These conditions were accepted by the Codex Alimentarius Commission in 1983 when drafting a general standard for irradiated foods. However, there has not yet been any marked expansion in commercial food irradiation, partly because of the perceived reluctance of consumers to accept irradiated foods.

FURTHER READING

Akamine, E.K.; Moy, J.H. Delay in postharvest ripening and senescence of fruits. Josephson, E.S.; Peterson, M.S. eds. Preservation of food by ionizing radiation. Vol. 3. Boca Raton, FL: CRC Press; 1983: 129–58.

American Society of Heating, Refrigeration and Air-Conditioning Engineers. ASHRAE Handbook of Refrigeration Systems and Applications. Atlanta, GA; 1986.

Commonwealth Institute of Entomology. Distribution maps of pests. London.

Cutter, E.G. Structure and development of the potato plant. Harris, P.M., ed. The potato crop. London: Chapman & Hall; 1978.

Diehl, J.F. Experiences with irradiation of potatoes and onions. Lebensm. Wiss + Technol. 10: 178–81; 1977.

Hardenburg, R.E.; Watada, A.E.; Wang, C.Y. The commercial storage of fruits, vegetables and florist and nursery stocks. Revised ed. Washington, DC: US Department of Agriculture; 1986. Agriculture Handbook no. 66.

Hill, A.R.; Rigney, C.J.: Sproul, A.N. Cold storage of oranges as a disinfestation treatment against fruit flies *Dacus tryoni* (Froggatt) and *Ceratitis capitata* (Wiedemann) (Diptera: Tephritidae). J.Econ. Entomol. 81: 257–60; 1988.

McCornack, A.A.; Wardowski, W.F. Degreening Florida citrus fruit: procedures and physiology. Proc. Int. Soc. Citriculture 1: 211–15; 1977.

McGlasson, W.B.; Kavanagh, E.E.; Beattie, B.B. Ripening tomatoes with ethylene using the trickle system. New South Wales Department of Agriculture, Sydney; 1986. Agfact H8.4.6.

Maxie, E.C.; Sommer, N.F.; Mitchell, F.G. Infeasibility of irradiating fresh fruits and vegetables, HortScience 6: 202–4; 1971.

May, J.H. Potential of gamma irradiation of fruits: a review. J. Food Technol. 12: 449–57; 1977.

Morris, S.C. The practical and economic benefits of ionising radiation for the postharvest treatment of fruit and vegetables: an evaluation. Food Technol. Aust. 39: 336–41; 1987.

Sawyer, R.L. Sprout inhibition. Talburt, W.F.; Smith, O., eds. Potato processing. 3d ed. Westport, CT: AVI; 1975: 157–69.

Sherman, M. Control of ethylene in the postharvest environment. HortScience 20: 57–60; 1985.

Soule, J. Citrus maturity and packinghouse procedures. University of Florida; 1974.

Thomas, P. Radiation preservation of foods of plant origin. I. Potatoes and other tuber crops. CRC Crit. Rev. Food Sci. Nutr. 19: 327–79; 1984.

Idem. II. Onions and other bulb crops. CRC Crit. Rev. Food Sci. Nutr. 21: 95–136; 1984.

Idem. III. Tropical fruits: bananas, mangoes and papayas. CRC Crit. Rev. Food Sci. Nutr. 23: 147–205; 1986.

Idem. IV. Subtropical fruits: citrus, grapes, and avocadoes. CRC Crit. Rev. Food Sci. Nutr. 24:53–89; 1986.

Idem. V. Temperate fruits: pome fruits, stone fruits, and berries. CRC Crit. Rev. Food Sci. Nutr. 24: 357–400; 1986.

Idem. VI. Mushrooms, tomatoes, monor fruits and vegetables, dried fruits, and nuts. CRC Crit. Rev. Food Sci. Nutr. 26: 313–58; 1988.

Urbain, W.M. Food irradiation. Orlando; Academic Press; 1986.

US Department of Agriculture. Plant protection and quarantine treatment manual, revised ed. Washington, DC: 1976.

World Health Organization. Wholesomeness of irradiated food. Report of the joint FAO/IAEA/WHO Expert Committee. Geneva, 1977 and 1981; WHO Technical Report Series Nos. 604 and 659.

11
Handling, packaging and distribution

Centres of consumption of fresh produce are usually remote from production areas. The costs of distribution in both money and energy terms, which includes handling, packaging and transportation, often exceed those of production. Careful management of the distribution system will ensure that produce retains its quality and that economic returns are maximized. Packaging is a key component of this system. Packaging has been practised as long as fresh produce has been traded. The two main functions of packaging are: 1. to assemble the produce into convenient units for handling; and 2. to protect the produce during distribution, storage and marketing. The earliest packages were mostly constructed of plant materials, such as woven leaves, reeds and grass stems and were designed to be carried by hand (Figure 47). Most packages are handled manually at some stage, and are sized accordingly. Often, however, they are assembled into larger units (such as onto pallets) for mechanical handling. Today, produce is transported and sold in a wide range of packages constructed of wood, fibreboard, jute (hessian), or plastics (Figures 48, 49, 50). Current methods of packaging are often wasteful of raw materials, as few packages are re-used.

Modern packaging for fresh produce is expected to meet a wide range of requirements which may be summarized as follows:

1. The packages must have sufficient mechanical strength to protect the contents during handling, during transport, and while stacked.
2. The material of construction must not contain chemicals which could transfer to the produce and be toxic to it or to humans.
3. The package must meet handling and marketing requirements in terms of weight, size, and shape. The current trend is to reduce the many sizes and shapes of packages by standardization. Palletizing and mechanical handling make standardization essential for economical operation.

Figure 47 Shoulder transport of taro in bamboo baskets in Indonesia.

4. The packages should allow rapid cooling of the contents. Furthermore, the permeability of plastic films to respiratory gases might also be important.
5. The mechanical strength should be largely unaffected by its moisture content when wet or at high humidities. The package might be required to exclude water from the produce, or to prevent dehydration of the produce.
6. The security of the package or its ease of opening and closing might be important in some marketing situations.
7. The package should identify its contents.
8. The package might be required to either exclude light or to be transparent.
9. The package might be required to aid retail presentation.
10. The package might need to be designed for ease of disposal, re-use, or recycling.
11. The cost of the package in relation to the value and the required extent of protection of the contents should be as low as possible.

EFFECT OF PACKAGING ON PRODUCE QUALITY

Prevention of mechanical damage

Fruit and vegetables vary widely in their susceptibility to mechanical damage and in the types of mechanical injury to which they are susceptible. The choice of package and packing method must take account of these differences. Four

Figure 48   A range of packaging materials and types normally encountered in the handling and marketing of produce in developed countries.

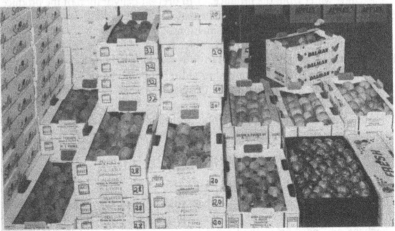

Figure 49 Interlocking polystyrene crates for produce.

Figure 50 Returnable plastic crates. Left, handles folded down for 'nesting' of empty crates; Right, handles in position for vertical stacking of filled crates. (Courtesy of W.E. Rushton. CSIRO Division of Food Processing.)

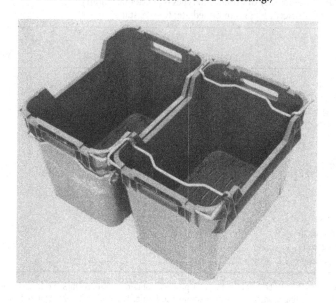

different causes of mechanical injury to produce can be delineated: cuts, compressions, impacts, and vibrations (rubs). The approximate susceptibility of some fruits to compression, impact and vibration injuries is shown in Table 28.

The avoidance of cuts is obvious. With the partial exception of certain, so-called, 'hard' vegetables such as pumpkin, the package must be strong enough to carry the stacking loads, otherwise there will be varying degrees of compression bruising of the contents. Impact bruising is caused by dropping of packages, or by impact shocks during transport. Bruising is tissue damage which results from strain energy being dissipated in the tissue and the amount of damage depends on how much energy is dissipated and the nature of the tissue. Vibration, common during transport, causes abrasion ranging from light rub marks to removal of, not only skin, but also some of the flesh. All these injuries turn brown through oxidation of tannins and exposure of similar materials to air in the damaged tissues; thus the produce is disfigured and its market value is reduced. Furthermore, these injuries are avenues for infection, increase respiration and hence the rate of deterioration, and cause immediate loss of edible material because of the need to trim off damaged portions.

Two important principles should be observed when packaging perishable produce: individual specimens should not move with respect to each other or the walls of the package to avoid vibration injury, and the package should be full without overfilling, or packing too tightly, which increases compression and impact bruising. Thus the individual items must be held firmly, but not too

Table 28 Susceptibility of produce to types of mechanical injury[1]
S—susceptible; I—intermediate; R—resistant

Produce	Compression	Impact	Type of injury Vibration
Apple	S	S	I
Apricot	I	I	S
Banana, green	I	I	S
Banana, ripe	S	S	S
Cantaloupe	S	I	I
Grape	R	I	S
Nectarine	I	I	S
Peach	S	S	S
Pear	R	I	S
Plum	R	R	S
Strawberry	S	I	R
Summer squash	I	S	S
Tomato, green	S	I	I
Tomato, pink	S	S	I

[1] From Guillou. R. Orderly development of produce containers. Proceedings. Fruit and Vegetable Perishables Handling Conference; University of California, Davis CA: 23–25 March 1964; 20–5.

tightly, within the package. In this respect packaging can be made more protective: by individually wrapping each piece of produce, or by isolating each piece, as in the cell pack and the tray pack methods for fruit, or by using energy-absorbing, cushioning pads. These techniques, however, increase costs which must be justified by reduced waste or increased selling price. Careful handling of the packages is the best defence against damage.

Mechanical strength of the package

For continued protection of produce against mechanical damage, the package must retain its strength throughout the marketing chain. Under conditions of high humidity, after condensation, or after being wet by rain, either the strength of the package must be independent of moisture content, for example, wood and jute, or the package material must not absorb moisture. The widely used fibreboard cartons and trays rapidly lose strength as they absorb moisture and so are less satisfactory in tropical conditions and in high humidity cool storage. They can, however, be protected if fully impregnated with wax or similar material, but wax impregnation is expensive. The corrugated fibreboard used in cartons is often only given a surface coating, which affords some protection from exposure to free water or to high humidities of short duration. Plastic films and plastic crates have the great advantage of absorbing little moisture.

Cooling the produce in the package

A further important requirement of packaging is that it must allow rapid cooling of produce. Both the nature of the produce and the treatment after packing must be considered. Respiratory heat must be able to escape readily to the surface of the package. In all individually wrapped commodities (whether in wooden boxes or cartons), and in baskets, sacks, or unvented packs of small commodities, such as green beans, peas, small fruits and leafy vegetables that naturally pack tightly, heat is removed solely by conduction to the surface of the package. The thickness of the mass of the contents (i.e. the minimum dimension of the package) becomes critical, and the acceptable thickness depends on the respiration rate of the commodity. If this thickness is exceeded, produce near the centre of the package will heat up because respiratory heat is not dissipated fast enough. In practice, self-heating becomes significant for rapidly respiring produce such as peas, beans, lettuce and broccoli in large single packages or close stacks of packages; such a problem may be avoided with smaller packages or by ventilating large packages or stacks.

Bulk bins that hold about 500 kilograms of produce are now widely used for harvesting, storage and often transport of apples, pears and other fruit and vegetables. Adequate cooling of fruit in stacks of bins is achieved in cool stores without forced movement of air, through a 'chimney effect', if about 10 per cent of the floor area of these bins is vented, but fan-assisted ventilation is necessary for rapid overnight cooling.

Effect of packaging on weight loss

Packaging generally minimizes weight loss and shrinkage of produce during marketing by the following techniques: wrapping or bagging individual items, or of small units of several items, in such materials as plastic film or waxed paper; fully lining the package with a material impervious to water vapour; increasing the resistance of fibreboard packages with a surface coating or by fully waxing the carton.

Under dry conditions, wooden boxes and crates, or baskets of produce may be deliberately sprayed with water; such direct wetting also assists in cooling produce by evaporation of the added water.

PACKING AND STOWAGE

Package dimensions

The dimensions of the package are important: the size and shape should provide economic use of materials, adequate strength, and easy and secure handling, loading and stacking. The optimal length width ratio is about 1.5:1. With the advent of mechanical handling of groups of packages there have been strong moves to standardize package sizes by selecting a smaller number that stack well on standard pallets (Figure 51). Smaller packages are also being used because of recommendations by the International Labour Organization about the maximum weights a person can reasonably be expected to handle routinely. A 30-litre package (holding about 20 kilograms of produce) and a 15-litre package are

Figure 51   Mechanical handling of fibreboard cartons stacked on wooden pallet.

becoming the standard packages for fruit, with a larger 36-litre package for some vegetables.

Standardization of package sizes promotes more efficient materials handling. This will require the produce to be fitted to the package rather than the package to the produce as in the past. The traditional baskets (Figure 52) so widely used in many developing countries, because they are relatively cheap, do not stack efficiently and are also often too large. Produce is damaged during handling and transport, and heat of respiration accumulates in the large masses of produce.

Packing

The desirable pack for most fruits and vegetables is one in which the package is tightly filled without bulging or overfilling so that the package, not the produce, bears the stacking loads. Some produce, such as potato, onion, root vegetables, and some citrus fruits, will withstand reasonable compressive loads and can be satisfactorily packaged in non-rigid packages such as sacks if they are carefully handled.

In developed countries, for many years fruit has been place-packed or pattern-packed, each piece being put into position by hand according to a pattern developed to maximize nett weight, to maintain a tight pack, and to present the fruit attractively when the package is opened. This time-consuming operation requires accurate grading into a range of size-classes, and is becoming too costly. The alternative is volume-fill or tight-fill packing, in which the fruit is poured into the carton or box. After filling, the pack is vibrated to secure a firm and tight pack. By this method produce is packed to a standard weight rather than a standard count. In a tight-filled pack, the produce is maintained firmly positioned with a pressure pad over which a lid is securely fastened. Package inserts, such as moulded pulp or plastic trays to isolate individual fruits, are expensive, but remain in use for delicate, high value produce. Individual wrapping still has a place, as does lining the package, to reduce vibration damage, moisture loss or both.

It is not practicable to recommend specific packages for each fruit or vegetable as several types may be satisfactory. The most suitable package would depend on many factors, such as the region, the length and nature of the market chain, methods of handling and transport, environmental conditions, availability and cost of materials and whether the produce is to be refrigerated. In many tropical countries the traditional basket made from bamboo (Figure 52) or other readily available materials is the cheapest, and often is adequate.

Stowage

Stowage or stacking of packages must simultaneously ensure stability of the stack and also provide for adequate air movement to enable satisfactory cooling and to remove respiratory heat. Stowage must also be economical of space and easy to achieve. A much stronger package is needed for long distance transport, or for high stacking in cool stores than for local marketing. Where packages are normally handled on pallets, the package must economically fit the pallet. Stack

Figure 52 Typical bamboo basket used throughout South-East Asia for handling and transport of produce.

stability is usually best obtained by either cross-stacking or tied-stacking, which require the packages to have suitable length to breadth ratios. Rounded baskets are more difficult to stack securely.

Some open stowage is necessary in any stack for cooling. An exception is where warming of previously cooled produce is to be restricted, when the stow should be as tight as possible. Tight stows, which naturally have a greater stability, are safe for short periods during which self-heating by respiration will not be a problem.

TESTING OF PACKAGES

Testing the performance of packages is an essential part of package research and development. The impacts and vibrations that occur during commercial handling and transport of packaged produce are complex and difficult to simulate in the laboratory. Useful laboratory tests include machine testing for compressive strength under standard conditions and simple drop tests. Quite complex transport shock simulators, computer programmed from actual transport operation records using strain gauges and so on, have been developed, but they are expensive and only worth-while for specialized research. There is no complete substitute for monitoring package performance under actual commercial conditions.

Figure 53 Fresh produce displayed in a modern self-service store. (Courtesy W.E. Rushton, CSIRO Division of Food Processing.)

CONSUMER PRE-PACKAGING

Produce purchased at retail outlets was traditionally packaged in paper bags, but paper bags have been largely replaced by bags of polyethylene film, which are cheaper, stronger, and transparent.

Supermarket handling accelerated the trend to marketing of pre-packaged fruit and vegetables where produce is pre-weighed and packaged into small units for retail sale. However, consumers have often been doubtful of the quality of prepackaged items and many countries have recently moved away from pre-packaging to allow customers to select their own produce from an open display.

The pre-packaging of produce in consumer-sized packages with plastic film or plastic or paperboard trays overwrapped with clear film (Figure 53), restricts weight loss from produce and can also provide a degree of modified atmosphere benefit. This latter characteristic is risky and generally not sought, so that films with low gas permeabilities are perforated to prevent significant modification of the package atmosphere. The number of small perforations necessary does not increase water loss appreciably.

Plastic films for packaging produce have good tensile strength under all likely environmental conditions and are inexpensive. Other properties, such as gas and water permeability, suitability for heat sealing, clarity and printability, can often be specified by the packer, as the manufacturer can develop a film to most required specifications. The film most widely used for pre-packaging fruit and vegetables is low density polyethylene. In addition to the general properties of plastics, polyethylene can be heat sealed, is usable over a wide temperature range ($-50°$ to $-70°C$), suits machine packaging lines, and is inexpensive, probably being the cheapest film in most countries. Polyethylene is permeable to many volatile compounds and gases but relatively impermeable to water vapour. The gas permeability can be controlled by varying either the density of the film or its thickness, or the film may be perforated.

FURTHER READING

Debney, H.G.; Blacker, K.J.; Redding, B.J.; Watkins, J.B. Handling and storage practices for fresh fruits and vegetables – product manual. Brisbane, Australia: Australian United Fresh Fruit and Vegetable Association (C/O Committee of Direction of Fruit Marketing); 1980.

Holt, J.E.; Schoorl, D. Fractures in potatoes and apples. J. Materials Sci. 18: 2017–28; 1983.

Holt, J.E.; Schoorl, D. Package protection and energy dissipation in apple packs. Sci. Hortic. 24: 165–64; 1984.

Holt, J.E.; Schoorl, D.; Lucas, C. Prediction of bruising in impacted multilayered apple packs. Transactions ASAE. 24: 242–7; 1981.

Kader, A.A. et al. Postharvest technology of horticultural crops. Berkeley, CA: University of California; 1985. Special Publication 3311.

Meffert, H.F. Th.; Beek, G. van. Safe radius is an important thermophysical property. Bull. Int. Inst. Refrig.; 1976. Annex 1976–1. 341–8.

O'Brien, M.; Cargill, B.F.; Fridley, R.B. Principles & practices for harvesting & handling fruits & nuts. Westport, CT: AVI; 1983.

Peleg, K. Produce handling, packaging and distribution. Westport, CT: AVI; 1985.

Schoorl, D.; Holt, J.E. A methodology for the management of quality in horticultural distribution. Agr. Systems. 16: 199–216; 1985.

12
Technology of storage

It has long been known that the preservation of fresh fruit and vegetables is governed by three major parameters:
(1) they keep better when cold; (2) they are damaged by freezing; and (3) they shrivel or wilt in dry air. On the basis of this knowledge, control of the temperature and humidity of the air around produce has been practised with increasing sophistication and success. A fourth parameter, the composition of the atmosphere, has become known only comparatively recently so that controlled atmosphere storage is a modern innovation.

METHODS OF STORAGE

In-ground storage
Pit storage, or clamp storage, is still practised in some developing countries. Hard vegetables, such as potato, turnip and late season cabbage, are piled into pits dug into a hillside or other well-drained spot. The pits are lined with hay or straw; the produce is then covered with straw followed by 10 to 20 centimetres of sods and earth to protect it against freezing and to shed rain. Piped ventilation to the outside, to reduce respiratory self-heating, is an advantage. Clamp storage is still a useful method of simple farm storage and has been shown to be suitable for storage of cassava for up to two months in the tropics (Figure 54). In Europe, perishable produce was initially stored in cellars and caves which, being below ground, were cooler than above-ground buildings in warm weather, and warmer in winter, reducing the risk of freezing. These methods are still commonly practised in the People's Republic of China.

Cellars are more sophisticated forms of below-ground storage; they may be part of above-ground buildings or underground rooms, often in hillsides, where access is easier. Again, good drainage and protection from rain are essential. The performance of cellars is improved by providing controlled ventilation openings

Figure 54 Cassava storage clamp. (Courtesy of R.H. Booth, Food and Agricultural Organization of the United Nations, Rome.)

for entrance of cold air and exit of warm air by convectional circulation when cooling is required. Although temperatures generally are not optimal, a good cellar will provide satisfactory storage for hard vegetables and long-keeping fruits such as apple.

Air-cooled stores
These are simply insulated structures above ground, or partly underground, which are cooled by circulation of colder, outside air. When the temperature of the produce is above the desired level, and if the temperature of the outside air is lower, air is circulated through the store by convectional or mechanical means through bottom inlet vents and top outlets fitted with dampers. Fans may be installed and are controlled manually, or automatically with differential thermostats. The air may be humidified, a process that can also be automated. Air-cooled stores are cheap to construct and to operate and are still widely used for the storage of potato and sweet potato, both of which need relatively high storage temperatures to avoid accumulation of sugar and chilling injury respectively.

Potatoes are commonly stord in bulk piles in stores with air delivery ducts under the floor, or at floor level, and with suitably spaced air outlets.

Ice refrigeration

An advance on air-cooled storage was the use of natural ice as a refrigerant. The lower temperatures obtained enabled longer storage of meat and other perishable foods, including horticultural produce. In North America, and North and Central Europe, ice was harvested in the winter from frozen lakes and ponds and stored in insulated 'ice houses'. The melting of 1 kilogram of ice absorbs 325 kilojoules, but the considerable bulk of ice needed and disposal of the melt water are disadvantages. The introduction of the small 'ice box' or 'ice chest' was a great advance in the domestic and small-scale commercial preservation of perishable foodstuffs.

Mechanical refrigeration

The father of modern refrigeration was undoubtedly the Australian James Harrison. By 1851 he had designed and built the first ice-making plant in the world incorporating a small refrigeration compressor with its auxiliary equipment and ice tank at Geelong in Victoria. In 1854 Harrison was granted British Patent No. 717 for 'the production of cold by the evaporation of volatile liquids in vacuo', an invention probably equal in importance to that of the steam engine. The general principles of Harrison's design remain virtually unchanged in modern refrigeration plants. The system was developed rapidly, and mechanically refrigerated cold stores, insulated with natural materials, such as sawdust or cork, were operating within a few years. A shipment of frozen beef from Australia to England in 1879 was the first, successful, long distance shipment of perishable food by sea; soon after, the first mechanically refrigerated cool stores for apple and pear were in operation.

A refrigeration plant consists of four basic components: the compressor in which the refrigerant gas, either ammonia or halogenated hydrocarbons, is compressed (and unavoidably heated); the condenser, either air-cooled or water-cooled, in which the hot gas is cooled and condensed to a liquid; the expansion valve; and the evaporator coils in which the liquid is permitted to boil and so remove heat from its surroundings (Figure 55). Fans are usually necessary to circulate the storage air over the cooling coils of the evaporator and through the stacks of produce in the store. The main agent for transfer of heat from the interior of the store to the cooling coils is air movement: radiation and convection may play a small part. In addition to these four basic components of a refrigeration plant and fans or correctly placed air ducts, several other items are required: various items of ancillary control equipment, a liquid receiver, and some means of defrosting the coils.

DESIGN AND CONSTRUCTION OF COOL STORES

A cool store is a large, thermally insulated box, with doors for entry and some means of cooling the interior. Cool stores for fruit and vegetables have special

Figure 55   Basic component parts of a mechanical refrigeration plant. (Adapted from Patchen, G.O. Storage for apples and pears. Washington, DC: US Department of Agriculture; 1971. Marketing Research Report No. 924.)

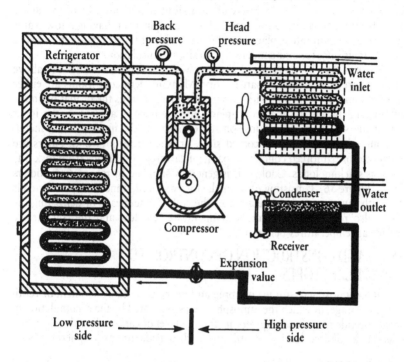

requirements in comparison with other refrigerated stores. These include a high cooling capacity, close control of temperature, and a relative humidity of 90 to 95 per cent. A common minimum design criterion is to provide capacity to cool a daily intake of 10 per cent of the capacity of the store at an initial rate of not less than 0.5 deg. C per hour. Such a capacity requires 1 tonne (3.5 kilowatts) of refrigeration capacity per 18 tonnes of produce for small stores of up to about 150 tonnes capacity, and about 1 tonne per 25 tonnes for larger stores. The capacity for larger stores can be varied by having two or more compressors or by the technique of cylinder unloading in one compressor.

Good temperature control requires spatial variation of no more than ±1 deg. C and a variation in time in any one position of no more than ±0.5 deg. C. A temperature difference of 1 deg. C over the storage period has significant effects on most produce, especially those stored at less than 5°C. The optimum thickness of insulation in walls and ceiling is the equivalent to 4 centimetres of cork per 10 degrees C difference in temperature between the inside and outside. This will keep the overall heat transfer to about 0.3 kilojoule per square metre per hour. This gives about the most economical ratio of cost of refrigeration capacity to cost of insulation and also enables maintenance of high humidities. The best insulation is the cheapest that gives the required performance; it may be 6 centimetres thickness of polyurethane foam or 40 centimetres of sawdust. Floors generally require half the thickness of insulation that is used in the walls. A vapour barrier of laminated foil material, or the equivalent, having a low water vapour transmission rate is placed on the warm side of the insulation, to prevent moisture migrating to, and condensing within, the insulation.

Cool stores may be constructed in many ways, and provided that the above conditions are met, all can be satisfactory. Many modern cool rooms are either sandwich panel construction with polystyrene foam slabs as the insulation in the prefabricated panels, or foamed-in-place polyurethane is applied to the inner faces of the structure. The 'skins' on the outsides of the insulation are metal, commonly aluminium or zinc-coated steel, or waterproof plywood. Floors are constructed of reinforced concrete capable of carrying point loads from fork lifts as well as stacking loads. Cooled air is generally supplied by forced draft coolers (FDCs) (Figure 56), consisting of framed, closely spaced, finned, evaporator coils fitted with fans to circulate the air over the coils. Some means of defrosting the coils is also required when storage temperatures are low and the coil surface operates at temperatures below 0°C.

DESIGN AND CONSTRUCTION OF CONTROLLED ATMOSPHERE STORES

Controlled atmosphere storage of apple and pear was initially a system of ventilated gas storage, in which the atmosphere was generated by the accumulation of carbon dioxide from fruit respiration, and the level of carbon dioxide was maintained at the desired level by ventilation with outside air. Atmospheres of such

Figure 56   Modern cool room construction of stressed-skin panels with polyurethane insulation. Cooled air is supplied by overhead FDCs.

stores contained 5 to 10 per cent carbon dioxide and 16 to 11 per cent oxygen as one volume of carbon dioxide is produced by the fruit for each volume of oxygen consumed. Further research revealed that some important cultivars were damaged by carbon dioxide concentrations above 3 per cent and that significant benefits could be obtained if store oxygen concentrations were in the range of 2 to 3 per cent. An atmosphere of 2 to 3 per cent carbon dioxide with 2 to 3 per cent oxygen was, therefore, found to be suitable for most cultivars of apple and pear at cool storage temperatures (Chapter 6). To be able to maintain such a low oxygen atmosphere, a much more gas-tight room was required. This required highly specialized methods of construction. Furthermore, ventilation of the store with outside air to control the carbon dioxide concentration was not possible as it would introduce too much oxygen; therefore, some means of absorbing, or 'scrubbing out', the excess carbon dioxide was required. Early carbon dioxide scrubbers relied on chemical absorption of the carbon dioxide in alkaline solutions, such as potassium hydroxide or calcium hydroxide. Later, less cumbersome and less messy methods were developed, by which carbon dioxide was adsorbed physically or absorbed chemically with dry hydrated lime.

Thus the controlled atmosphere store has to be relatively gas-tight, and fitted with a means of measuring and controlling the concentrations of both carbon dioxide and oxygen. Being a sealed chamber, the refrigeration system has to be completely reliable, and the room has to be fitted with adequate, accurate and reliable, remote-reading thermometers.

A more recent development, of considerable practical significance, is the external generator (Figure 57) which consumes the oxygen in the air much more rapidly than the produce can by respiration. The generators is a special burner, operating on gaseous fuel, which either produces a low oxygen atmosphere with the required carbon dioxide content to flush out the oxygen from the room (flushing system) or consumes the oxygen from the air in the room itself (recirculating system). A carbon dioxide adsorber is also required to adsorb the excess carbon dioxide produced by the generator and by the fruit. Such a generator will also enable a 2 to 3 per cent oxygen atmosphere to be maintained in a relatively leaky room. Alternatively, low oxygen atmospheres can be achieved by flushing the room with liquid or compressed nitrogen, or with an atmosphere containing low oxygen from gas separators such as the pressure swing adsorption or hollow fibre membrane systems. The development of generators and the now relatively simple operation of low-oxygen stores has probably been largely responsible for the current popularity of controlled atmosphere storage. The value of reducing the oxygen concentration rapidly by artificial means has been in dispute. But the value of building a gas-tight store, in which a reasonable rate of oxygen reduction can be achieved without a generator, and with which a generator can be used more economically, has never been doubted.

Construction

An essential feature of a controlled atmosphere store is the provision of an effective gas barrier, which is most conveniently placed directly on the inside of the

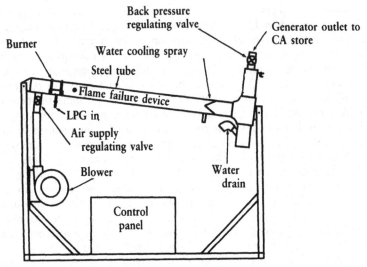

Figure 57   Diagram of an open-flame generator. Air is drawn from the atmosphere, mixed
with liquefied petroleum gas and burnt:

$$CH_4 + 2O_2 \leftarrow CO_2 + 2H_2O$$

The combustion gases are cooled with the water spray and blown into the controlled atmosphere store. Commercial equipment includes control and safety
devices to shut off gas supply in the event of failure of components. (From
Holligan, P.J.; Scott, K.J. Controlled atmosphere storage using a plastic room
and an open-flame generator—a manual for operators. Sydney: Department of
Agriculture NSW; 1973. Technical Report No. 2/73. With permission.)

insulated surface. If the external vapour barrier is defective, however, moisture
that penetrates this barrier will then be contained on the inside of the gas barrier,
leading to water-logging and destruction of the insulation. The concept of the
'jacketed' room overcomes this and allows the gas barrier to remain readily
accessible. The system provides a gas-tight lining inside a cool room, with an air
space between the insulated walls and the lining. Cold air is circulated through
this narrow space to remove the heat. An expensive disadvantage of the system
is the need for under-floor air ducts. The 'blanket' type of store is a variation of
the jacket type, and has a normal floor which reduces the cost of construction
(Figure 58). Primary cooling air circulates over the ceiling and around the walls.
A further variation is a room in which a plastic film acts as a blanket, that is,
'plastic tent' controlled atmosphere storage. Such rooms have operated satisfactorily in Australia and New Zealand. Rigid structures, besides being difficult
to seal initially, are prone to develop leaks because of varying pressure differentials between the outside and inside due to changes in atmospheric pressure and
the normal cycling of the refrigeration system. This problem is partially overcome
by jacket or blanket construction and completely overcome with the plastic tent.

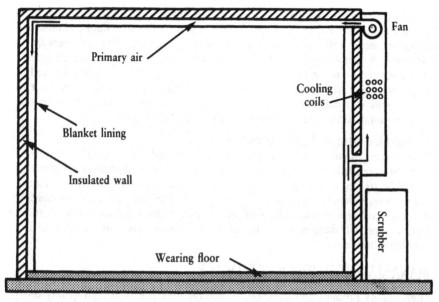

Figure 58 Section of a blanket cool store. (From Atkins, R. Controlled-atmosphere stores for fruit. The blanket system. CSIRO Food Res. Q. 33: 75–80: 1973. With permission.)

The tent is flexible and automatically varies the internal volume as pressure differentials develop thus avoiding significant pressure effects.

A recent development, which enables the satisfactory conversion of existing cool stores to controlled atmosphere operation and also cheap new construction, is the complete, internal lining of a simple metal chamber with foamed-in-place polyurethane. If correctly applied, the polyurethane provides both insulation and a gas barrier. A pressure relief device, usually a water trap, is fitted through the walls of such a rigid, gas-tight structure to avoid damage by limiting pressure differentials to 370 pascals (15 millimetres water gauge).

Types of generators
Flushing generators burn liquefied petroleum gas (LPG) to remove oxygen from the air and purge the room continuously with the resultant gas. A recirculating generator operates similarly except that the atmosphere in the room is recycled through the generator. Factors to be considered when choosing between flushing and recirculating generators are capacity, room gas-tightness, capital and operating costs. The capital cost of a recirculating generator is about three times that of the simplest open-flame, flushing generator, but if the room is gas-tight the consumption of LPG will be about one half and if leaky, many times more. The open-flame generator is more economical for blanket-type rooms which are gas-tight and readily maintained.

The excess carbon dioxide generated can be removed by commercial scrubbers, which employ either reagents, such as sodium hydroxide, potassium carbonate, ethanolamine or water, or adsorption onto activated carbon. These scrubbers are expensive. A scrubber utilizing dry hydrated lime removes excess carbon dioxide effectively and is easy and cheap to operate. The lime is held in paper bags stacked on pallets which are loaded into an external scrubber chamber, made from either a metal tank or a plastic tent sealed to an impervious floor. Annual requirement is approximately 500 grams of lime per 20 kilograms of apples.

Controlled atmosphere stores are lethal

Atmospheres in controlled atmosphere stores, although they support plant life at a low level, will not support mammalian life. Controlled atmosphere stores should be treated with respect to ensure that no one is ever exposed to such an atmosphere, unless wearing an efficient respirator with its own oxygen supply. Transport vehicles, in which the atmosphere has been modified with liquid nitrogen, are especially dangerous and, like controlled atmosphere stores, must be well ventilated before entry.

MANAGEMENT OF PRODUCE STORAGE

High quality produce will come out of storage only if it is of high quality on entering the store, and if management of the storage facilities is of high standard. Given correct selection and handling of the fruit, the success of subsequent storage depends on: quickly reducing the temperature of the fruit to the desired level and maintaining it with little variation; close maintenance of the desired humidity and gas concentrations in the storage atmosphere; avoiding over-storage.

Precooling the rooms

Storage rooms are generally brought down to the appropriate temperature a few days before commencement of intake of fruit. Three days is enough for a fully insulated room, but rooms without floor insulation are precooled for a week to ensure that the floor has cooled down to equilibrium before loading. Failure to precool the room before loading is often the cause of unsatisfactory maintenance of temperature, slow cooling, and excessive shrinkage of the produce.

Temperature control

Air movement transfers heat from the fruit to the coils by natural convectional circulation in a room with overhead grids (cooling pipes), by forced circulation in rooms cooled by forced draft coolers or by a combination of natural and forced convection. It follows that the nature of the packages and the method of stacking must allow the air to move readily through all parts of the stack for the produce to be cooled quickly and uniformly.

Spatial variation in produce temperature in a good store should not exceed 1 degree C above or below the nominal storage temperature. Several factors influence the spatial distribution of temperature in a store. The most important single requirement for uniform produce temperatures is uniform cooling over the entire area on the top of the stack. This applies equally to the distribution of air from forced air circulation systems and to the even distribution of the coils over the ceiling in natural circulation rooms. Also of importance is the uniformity of the air paths through the stow as air always takes the path of least resistance. Ideally, there should be a continuous, narrow, air slot in the direction of air flow past at least two faces of every box or carton and each side of every bulk bin, together with no large vertical gaps in the stack to allow short-circuiting by the cool air. The room should be well insulated to reduce heat leakage, and the coolers should have ample capacity to ensure a small difference between the temperature of the air and coil surface.

Selection, sorting and handling of produce

It is desirable to sort and size-grade produce before storage. Not all commodities are fit for storage; some have better keeping qualities than others, some are blemished (and usually sold to processors), and some of the produce harvested is unmarketable. Refrigerated storage is expensive, and it is not economic to have produce which is not fit for sale, or produce which would be better marketed immediately, occupying cool storage space. Sorting and sizing before storage, so that both quantity and quality in storage are known, is of great assistance in orderly marketing.

Loading

If possible warm produce should be cooled in a separate cool room from that used for storage. If only one room is available, the designed daily intake (commonly 10 per cent of capacity) should not be exceeded. Otherwise, the life of the produce will be reduced and shrinkage promoted. Warm produce should be loose stacked, and cooling can be improved with the aid of an auxilliary, portable fan placed in front of the stack, with the suction side to the produce, to draw air through it (Chapter 4).

Stacking

It is bad practice to over fill a room, as it results in variable temperatures and therefore a poor outturn of a proportion of the produce. Packaged produce is carefully stacked to give economy of space, adequate and uniform air circulation, and accessibility. The following requirements for stacking are essential for rapid cooling and good temperature control for any type of package:

1. Keep the stack 8 centimetres away from outer walls and 10 to 12 centimetres away from any wall exposed to the sun. This will ensure that heat coming in through the walls will be carried away to the coils by air moving freely between the stack and the wall without warming nearby produce.
2. Leave a clear air space of not less than 20 centimetres between overhead coil drip trays and the top of the stack. If unit coolers or other forced air circulation systems are used, the clear space between the top of the stack and the ceiling is generally not to be less than 25 centimetres. This ensures that a uniform layer of cold air blankets the whole stack. The full depth of the space

in front of a forced draft cooler is kept clear for a distance of 2 metres to allow it to function properly and to avoid freezing produce.

3. An air plenum of about 8 centimetres is required between the floor and the stack. When bins and pallets of boxes are used, the pallet bases provide the necessary air gap above the floor. Wherever possible they are placed with the pallet bases parallel to the direction of airflow, that is, running towards the forced draft cooler.

4. Leave small, vertical air paths within the stack, not less than 1 centimetre wide between adjacent packages. Freely exposed cartons cool at a similar rate to that of packed boxes of the same dimensions. But cartons, having straight sides, require special treatment in stacks. A layer-reversed, open chimney, stacking pattern will provide the necessary vertical gaps between cartons (1 centimetre) and at the same time provide a stable stack.

5. Bulk storage bins should have air gaps in the floor of 8 to 10 per cent of the base area. Rapid cooling of produce is possible in such bins. Bins of warm produce are first stacked only two-high overnight to allow quick removal of field heat from the produce. Next day they may be stacked to full height. Unless high humidities are maintained in the cool store, produce in bins that also have air gaps in the side may shrivel excessively. The sides of the bins can be lined to reduce shrivel, but cooling will be slow if the slatted bottoms are lined. Around each column of bins, at least at the sides which are at right angles to the pallet-base bearers, vertical air gaps about 4 centimetres wide are left, as this allows free escape of the air rising by convection through the produce in the bin.

Weight loss and shrinkage

Excessive shrinkage is due to immaturity of the produce, delay before storage, picking produce when hot and placing hot produce in the cool store, packing produce into dry wooden boxes or cartons, high storage temperatures (e.g. hot spots in the room), low humidities due to insufficient insulation or insufficient coil surface, slow cooling, and excessive air circulation. Fast cooling, uniformly low temperatures, and high humidities in the store are, therefore, necessary for low weight-losses. The extra cost of additional cooling and insulation, and a good vapour barrier can be more than offset by reduced weight-loss and better produce condition after storage.

Weight loss during cooling may also be greatly reduced by wetting warm produce, such as leafy vegetables, before it is put into the store. It is preferable to harvest produce early in the morning, when it is coolest and to put it directly into the cool store. This reduces the load on the refrigeration plant and lowers costs. When it is necessary to harvest produce later in the day in hot, dry weather, it may be practicable to hose some types of produce with water, to leave it overnight in the open to cool by a combination of evaporative cooling and radiation cooling (if the night sky is clear), and to put it into store next morning.

Orderly marketing and over-storage

Over-storage is still one of the commonest faults in the cool storage of produce. It is good marketing practice to commence selling long-keeping produce, such as apple and pear, from the cool store early and to continue regularly throughout the season. To achieve orderly marketing, produce in the cool store needs to be segregated according to its expected keeping quality and removed for sale accordingly; as a general rule produce first in should be first out.

Over-storage of produce may be minimized by placing aside a few small units of the various lines of produce; these are removed to room temperature at intervals during storage and at the first sign of deterioration of the sample the whole line should be marketed without delay. A further important reason for making such regular inspection of samples during storage is that some fruits may look in fine condition in the cool store but may either develop physiological disorders or fail to ripen satisfactorily after removal.

Sanitation

Cool rooms should be thoroughly cleaned at the end of each season and, if necessary, sterilized to reduce the risk of losses by mould attack: the walls and floor can be washed with a solution of sodium hypochlorite (chlorine) followed by fumigation with formaldehyde gas. Mouldy or otherwise contaminated bins and boxes should be cleaned and sterilized with steam or a fungicide before reuse. Grading machines are often an important source of mould contamination that leads to development of rots in storage. These machines should be cleaned and swabbed or sprayed with a fungicidal solution daily. The equipment should be regularly inspected for defects and any points likely to cause produce injury should be repaired.

REFRIGERATED TRANSPORT

Much produce is transported over long distances on land and sea under refrigeration. Refrigerated road or rail vehicles can be regarded as insulated boxes fitted with modular mechanical refrigeration units powered from diesel units. Refrigerated ships have a central refrigerating plant; the whole ('reefer ships') or only part of the carrying space on a vessel may be insulated and refrigerated.

Much refrigerated sea freight is now carried in containers of 30 or 60 cubic metres capacity, which permit temperature control from door to door. One type of refrigerated freight container, the 'integral' container, has its own refrigeration unit, operated electrically, and perhaps also incorporates a diesel-powered generator. The other type, the 'porthole' container, is a passive unit, which must be supplied with cool air from a 'clip-on refrigeration unit or from a central unit (Figure 59).

Economy of space is a prime requirement in all transport; therefore, refrigerated transport vehicles and containers are designed for high density stowage. They are not designed for rapid cooling, so that successful refrigerated transport requires thorough precooling of the load. Respiratory heat is a significant proportion of the refrigeration load, consequently some air space must be provided between the packages during stowage of produce unless the journey is short. Significant amounts of heat enter refrigerated road transport vehicles from the outside air, solar radiation, heat reflected from the road and from air leakage through the doors. To ensure the maintenance of even temperatures it is necessary to provide

Figure 59   Insulated refrigerated shipping container and a clip-on refrigeration unit. The matching portholes in the container and clip-on unit provide for the circulation of cool air through the clip-on unit and the contents of the container. (Courtesy Acta Pty. Limited, Sydney, Australia.)

Figure 60   To ensure good circulation there must be an air delivery chute, ribs on the doors and walls, ribs or pallets on the floor and a return air bulkhead. (From Sharp, A.K.; Irving, A.R.; Beattie, B.B. Transporting fresh produce in refrigerated trucks. NSW Department of Agriculture, Sydney; 1985. Agfact H1.4.3.)

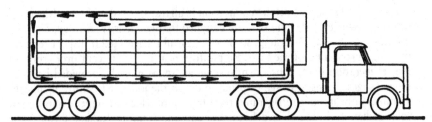

good circulation of cooling air around the load (Figure 60). Rules covering precooling, stowage, and air circulation have been developed from research and commercial experience; maximum acceptable loading temperatures are commonly specified and ought to be closely policed.

FURTHER READING

American Society of Heating, Refrigerating and Air-Conditioning Engineers. ASHRAE Handbook of Refrigeration Systems and Applications. Atlanta, GA; 1986.

Atkins, R. Fruit cool stores—the insulated structure. CSIRO Food Preserv. Q. 30: 51–5; 1970.

Blankenship S.M. ed. Controlled atmospheres for storage and transport of perishable agricultural commodities. Proceedings of the fourth national controlled atmosphere research conference, 23–26 July, 1985. Raleigh, NC: 1985. North Carolina State University Report No. 126.

Dalrymple, D.G. The development of an agricultural technology: controlled-atmosphere storage of fruit. Technol. Culture 10: 35–48; 1969.

Debney, H.G.; Blacker, B.J.; Redding, B.J.; Watkins, J.B. Handling and storage practices for fresh fruits and vegetables—produce manual. Brisbane, Australia: Australian United Fresh Fruit and Vegetable Association (C/O Committee of Direction of Fruit Marketing); 1980.

Food and Agriculture Organization of the United Nations. Refrigeration applications to fish, fruit and vegetables in South East Asia. Rome; 1974.

Holligan, P.J.; Scott, K.J. A prefabricated plastic room and open flame generator for the controlled atmosphere storage of apples. Food Technol. Aust. 23: 336–9; 1971.

International Institute of Refrigeration. Recommendations for chilled storage of perishable produce. Int. Inst. Refrig., Paris; 1979.

Irving, A.R. Code of practice for handling fresh fruit and vegetables in refrigerated shipping containers. Department of Primary Industries and Energy, Canberra; 1988.

Kasmire, R.F.; Hinsch, R.T. Factors affecting transit temperatures in truck shipments of fresh produce. Davis, CA: University of California; 1982. Perishables Handling, Transportation Bull. No. 1.

Appendix I

LIST OF ABBREVIATIONS

ADP	adenosine diphosphate
ATP	adenosine triphosphate
Ca	calcium
CA	controlled atmosphere
C_2H_4	ethylene
CIPC	3-chloroisopropyl-N-phenylcarbamate
CO	carbon monoxide
CO_2	carbon dioxide
EDB	ethylene dibromide
EMP	Embden-Meyerhof-Parnas
ERH	equilibrium relative humidity
FDC	forced draft cooler
HOPP	*ortho*-phenylphenol
LPG	liquefied petroleum gas
K	potassium
$KMnO_4$	potassium permanganate
MA	modified atmosphere
N_2	nitrogen
NAD	nicotinamide adenine dinucleotide
$NADH_2$	reduced NAD
O_2	oxygen
OPP	*ortho*-phenylphenate (free anion)
pH	log value of hydrogen ion concentration
Pi	inorganic phosphate
Q_{10}	temperature gradient (10 deg. C)
RH	relative humidity
RNA	ribonucleic acid
RQ	respiratory quotient
SO_2	sulphur dioxide
SOPP	sodium *ortho*-phenylphenate
TBZ	thiabendazole
TCA	tricarboxylic acid
VPD	vapour pressure deficit

Appendix II

GLOSSARY OF BOTANICAL NAMES

Common and botanical names of some fruits and vegetables

Common name	Botanical name
Apple	*Malus x domestica* Borkh.
Apricot	*Prunus armeniaca* L.
Asian pear	*Pyrus pyrifolia* Nakai *and P. bretschneideri* Rehder.
Asparagus	*Asparagus officinalis* L.
Avocado	*Persea americana* Mill.
Banana	*Musa* L. sp. Cavendish varieties *M. acuminata* Colla
Beans, broad	*Vicia faba* L.
string	*Phaseolus vulgaris* L.
mung	*Phaseolus aureus* Roxb.
Beet	*Beta vulgaris* L.
Blueberry	*Vaccinium* sp.
Broccoli	*Brassica oleracea* L. (Italica group)
Brussels sprout	*Brassica oleracea* L. (Gemmifera group)
Cabbage	*Brassica oleracea* L. (Capitata group)
Carambola	*Averrhoa Carambola* L.
Carrot	*Daucus carota* L.
Cassava (manioc, tapioca)	*Mannihot esculenta* Crantz
Cauliflower	*Brassica oleracea* L. (Botrytis group)
Celery	*Apium graveolens* L.
Cherimoya	*Annona cherimola* Mill.
Cherry, sweet	*Prunus avium* L.
sour	*Prunus cerasus* L.
Chili	*Capsicum annuum* L.
Choko	*Sechium edule* (Jacq.) Sw.
Corn (maize), sweet	*Zea mays* L.
Cucumber	*Cucumis sativus* L.
Eggplant (Aubergine)	*Solanum melongena* L.
Feijoa	*Feijoa sellowiana* Berg.
Fig	*Ficus carica* L.
Garlic	*Allium sativum* L.
Ginger	*Zingiber officinale* Rascoe

Globe artichoke	*Cynara scolymus* L.
Grape	*Vitis vinifera* L.
Grapefruit	*Citrus paradisi* Macfad.
Guava	*Psidium guajava* L.
Jackfruit	*Artocarpus heterophyllus* (Lam.) L.
Jerusalem artichoke	*Helianthus tuberosus* L.
Kiwi fruit (Chinese gooseberry)	*Actinidia deliciosa* (A. Chev.) C.F. Liang et A.R. Ferguson
Leek	*Allium ampeloprasum* L.
Lemon	*Citrus limon* (L.) Burm. f.
Lettuce	*Lactuca sativa* L.
Lime	*Citrus aurantifolia* (Christm.) Swingle
Litchi	*Litchi chinensis* Sonn.
Loquat	*Eriobotrya japonica* Lindl.
Mandarin	*Citrus reticulata* Blanco
Mango	*Mangifera indica* L.
Mangosteen	*Garcinia mangostana* L.
Muskmelon (cantaloupe, honey dew)	*Cucumis melo* L.
Nectarine	*Prunus persica* (L.) Batsch.
Okra	*Hibiscus esculentus* L.
Onion	*Allium cepa* L.
Papaya	*Carica papaya* L.
Parsley	*Petroselinum crispum* (Mill.) Nym.
Parsnip	*Pastinaca sativa* L.
Passionfruit	*Passiflora edulis* Sims
Pea	*Pisum sativum* L.
Peach	*Prunus persica* (L.) Batsch.
Pear	*Pyrus communis* L.
Pepino	*Solanum muricatum* Ait.
Peppers, green and red	*Capsicum annuum* L.
Persimmon	*Diospyros kaki* L.f.
Pineapple	*Ananas comosus* (L.) Merr.
Plum	*Prunus domestica* L.
Pomegranate	*Punica granatum* L.
Potato	*Solanum tuberosum* L.
Pumpkin	*Cucurbita pepo* L.
Radish	*Raphanus sativa* L.
Rambutan	*Nephelium lappaceum* L. var. *esculentum* Nees
Rhubarb	*Rheum* sp.
Satsuma mandarin	*Citrus unshu* Mari
Soya bean	*Glycine max* (L.) Merr.
Spinach. European	*Spinacia oleracea* L.
Squash	*Cucurbita maxima* Duch.
Strawberry	*Fragaria* x *ananassa* Duch.
Sweet orange	*Citrus sinensis* (L.) Osbeck
Swede turnip	*Brassica napus* L. (Napobrassica group)
Sweet potato	*Ipomea batatas* (L.) Lam.
Tamarillo (tree tomato)	*Cyphomandra betacea* (Cav.) Sendt.
Taro	*Colocasia esculenta* (L.) Schott
Tomato	*Lycopersicon esculentum* Mill.
Turnip	*Brassica campestris* L. (Rapifera group)
Watermelon	*Citrullus lanatus* (Thunb.) Mansf.
Yam	*Dioscorea batatas* Deene.

Appendix III

SYSTÉME INTERNATIONAL D'UNITÉS (SI)[1]

Conversion table for postharvest measurements

Quantity	Imperial unit	Metric unit	Conversion factors Imperial to metric units
Length	inch (in)	centimetre (cm)	1 in = 2.54 cm
	foot (ft)	centimetre (cm)	1 ft = 30.48 cm
	yard (yd)	metre (m)	1 yd = 0.914 m
Mass	ounce (oz)	gram (g)	1 oz = 28.35 g
	pound (1b)	kilogram (kg)	1 lb = 0.454 kg
	ton	tonne (t)	1 ton = 1.02 t
Area	square inch (in^2)	square centimetre (cm^2)	$1 \ in^2 = 6.45 \ cm^2$
	square yard (yd^2)	square metre (m^2)	$1 \ yd^2 = 0.836 \ m^2$
Volume	cubic inch (in^3)	cubic centimetre (cm^3)	$1 \ in^3 = 16.4 \ cm^3$
	cubic foot (ft^3)	cubic metre (m^3)	$1 \ ft^3 = 0.0283 \ m^3$
	cubic yard (yd^3)	cubic metre (m^3)	$1 \ yd^3 = 0.765 \ m^3$
	bushel (bus)	cubic metre (m^3)	1 bus = 0.0364 m^3
	fluid ounce (fl oz)	millilitre (mL)	1 fl oz = 28.4 mL
	gallon (gal UK)	litre (L)	1 gal = 4.546 L
	gallon (gal US)	litre (L)	1 gal = 3.785 L
Pressure	atmosphere (atm)	kilopascal (kPa)	1 atm = 101 kPa (760 mm Hg, 408 cm H_2O)
Temperature	degree Fahrenheit (°F)	degree Celsius (°C)	$C = \frac{5}{9} (°F - 32)$
Energy	calorie (cal)	joule (J)	1 cal = 4.1868 joule
	British thermal unit (Btu)	kilojoule (kJ)	1 Btu = 1.055 kJ

[1] Adapted from Style manual for authors, editors and printers of Australian Government publications. 3rd ed. Canberra: Australian Government Publishing Service: 1978.

Appendix IV

TEMPERATURE MEASUREMENT

The Celsius, or Centigrade, (°C) and the Fahrenheit (°F) scales are the main temperature scales used in the science and commerce of fruit and vegetables. The Celsius scale is based on phase changes of water, with the freezing point of water being 0°C and the boiling point of water at atmospheric pressure being 100°C. It is the approved SI temperature scale (Appendix III) and is, therefore, being used in an increasing number of countries. The Fahrenheit scale, which has been widely used, records the freezing and boiling points of water as 32°F and 212°F respectively.

Temperature measuring devices

There is a range of devices that can be used to measure temperature. All of these devices need to be used with care if correct temperatures are to be measured, and all need calibration.

Liquid-in-glass thermometers

These thermometers are the most commonly used temperature measuring devices and are based on the principle that a liquid expands when it is heated and contracts when cooled. The change in volume is read against a fixed scale. They are cheap, simple, easy-to-read instruments with an acceptable rate of response to changes in temperature and their calibration changes little with time. But they are fragile and must be handled with care; the sensing bulb or liquid reservoir is necessarily thin-walled and the *stem will not bend*—a common reason for breakage! Commonly used liquids are mercury and ethyl alcohol (coloured, usually red, for easy visibility); their freezing and boiling points are −38.9°C and 356.6°C for mercury and −115°C and 78.3°C for alcohol. Mercury is a poisonous substance and escaped mercury must be treated accordingly.

These thermometers will give incorrect readings if not properly constructed or used. Unless supplied with a standard calibration certificate, every thermometer should be checked at 0°C in an ice water mixture and against a calibrated thermometer near the temperature at which it is to be used. Liquid-in-glass thermometers are designed to be either fully immersed or partly immersed in the material whose temperature is being measured and are marked accordingly. When using thermometers care should be taken that either extraneous heat from a hot body, for example your own, or a cold surface does not affect the reading. Sufficient time must also be allowed for the thermometer to come to equilibrium as the glass has some heat capacity and conductance.

Fluid-filled dial thermometers

These thermometers enable measurements to be made at a distance of a few metres; therefore, they are often used as externally indicating, dial thermometers in cool stores. They consist of a sensing bulb connected by a capillary to a spiral tube, called a Bourdon tube, and the whole system is filled with a liquid, a gas, or a saturated vapour. The pressure changes in response to varying temperature cause movement of the Bourdon tube which is linked to a pointer that indicates the temperature on a graduated circular dial. These thermometers are constructed with small bulbs made from a material such as copper which has a low specific heat and is highly conductive. Response to changes in temperature is slower than that of liquid-in-glass thermometers. These thermometers should be calibrated annually.

Bimetallic thermometers

These thermometers comprise a laminate of two metals, one with a high and one with a low coefficient of expansion. The laminate changes shape with temperature variation and activates a pointer or dial; the response to changes in temperature is slow. The indicated reading is subject to small errors because of friction in the mechanical components. These errors can be partially overcome by gently tapping the dial. Bimetallic thermometers are robust and are convenient for measuring internal flesh temperatures of produce by insertion of the pointed measuring tip. To minimize errors by conduction of heat along the metal stem, the thermometer should be immersed to the depth indicated, or to at least twenty times the diameter of the stem.

Thermographs

A thermograph combines a bimetal thermometer with a moving paper chart to give a temperature record. The chart is driven by clockwork or electric battery.

Thermocouples

If two strips of different metals are joined together at each end and the junctions are kept at different temperatures, an electromotive force (emf) develops in the two-part conductor; this electromotive force depends on the metals and the temperature difference. This principle is used in most thermoelectric thermometers. The simplest form comprises the thermocouple of two wires soldered together at their ends with a meter in the circuit to measure the generated electromotive force. One junction is kept at a known temperature, commonly at 0°C in melting ice, and the other is the sensing element placed where the temperature is to be measured. A commonly used pair of metals is copper and constantan (an alloy of 60 per cent copper and 40 per cent nickel) and gives an electromotive force of about 39 microvolts per degree Celsius.

Thermocouples have the following advantages: the length of the wires is not significant so that they can be used in remote positions; the measuring probe can be small; high precision and rapid responses can be obtained. Readings can be taken manually or recorded automatically. Thermocouples are most useful for studying temperature changes in space and time, for example, in a filled cool store or a loaded transport vehicle.

Resistance thermometers

Resistance thermometers have many of the advantages of thermocouples. They can be used for remote measuring, are accurate, and do not require a cold junction. The sensor contains a temperature-sensitive wire, usually of platinum, or a thermistor. They are robust as the sensor can be protected, and the cable to the measuring device can be heavy and well insulated. Because of their robustness they are commonly used for permanent distant reading installations, for example, in refrigerated ships.

Digital thermometers

A digital thermometer uses an electronic circuit to detect the output of a thermocouple resistance element or thermistor, converts this to a temperature, and shows the reading on a digital display. Although the temperature may be displayed to a tenth of a degree, digital thermometers are not necessarily as accurate as liquid-in-glass thermometers. Both hand-held, battery-operated and panel-mounted, mains-powered digital thermometers are in use. The accuracy of hand-held thermometers used in the field should be checked frequently at 0°C.

Digital data-loggers

As an extension of the digital thermometer, the data-logger measures temperatures at preset intervals of time, and stores the values in solid-state memory for later recall and analysis. Battery and mains-powered models are available.

Calibration of thermometers

Melting ice, made from potable water, has a temperature within 0.05° of 0°C. For checking at 0° an ice water mixtures should be prepared in an insulated container e.g. 'Thermos' flask, rather than in a beaker or glass jar. Another method is to use wet crushed ice and to tamp the ice down gently in a shallow container to obtain a 'solid' mass. The surface of the ice should look wet. The stem of the thermometer to be tested should be gently forced into the ice. Alternatively, a thick ice and water slurry can be used but this should be well-stirred. For checking at other temperatures, a calibrated mercury-in-glass thermometer with 0.1°C graduations, should be obtained. A thermometer with a calibration certificate may be purchased or a thermometer may be submitted to an approved testing laboratory for calibration.

Where to measure temperature

The placement of temperature measuring devices in cool stores and similar structures is important, particularly when the rates of cooling (or warming) of produce are being followed or when refrigeration or heating is being controlled by thermostats. The nature of the refrigeration system means that there must always be a gradient in temperature between the produce and the cooling coils. In addition, the gradient is accentuated by excessive heat leakage through the structure of faulty stacking of packages. Generally the aim is to maintain the bulk of the stow at the recommended storage temperature, without freezing or over-cooling the coldest part. The device that is used as a thermostatic control for the forced draft cooler is best placed in the air off the cooler, and the thermostat adjusted to give the minimum acceptable temperature.

Temperature of produce in a cool store, ship's hold, or container should be measured in several different positions, because there will inevitably be spatial variations which may be enough to adversely affect the cooling of some of the load. Air temperatures just inside the door never give accurate readings of produce temperature.

FURTHER READING

Hall, J.A. Fundamentals of thermometry. London: Institute of Physics Monographs for Students: 1953.
Henry. Z.A., ed. Instrumentation and measurement for environmental sciences. St. Joseph, MI: American Society of Agricultural Engineers: 1975.
Middlehurst, J. Temperature measurement. CSIRO Food Preserv. Q. 24: 5–10: 1964.

Appendix V

HUMIDITY MEASUREMENT

The amount of water vapour in air, that is, the psychrometric state of the atmosphere, can be specified either by the water content or by its vapour pressure and either in absolute or relative terms.

Relative humidity (RH) is the ratio of water vapour pressure in air to saturation vapour pressure at the same temperature, expressed as a percentage.

$$RH = (P/P_0)T \times 100\%$$

where P = water vapour pressure of air at temperature T
P_0 = saturation vapour pressure at the same temperature T
It is important to remember that relative humidity can only be compared at the same temperature and barometric pressure.

Absolute (or specific) humidity is the measure of the weight of water vapour contained in a known weight of dry air. Typical psychrometric charts provide a scale showing absolute humidity in grams or kilograms of water vapour per kilogram of dry air. Absolute humidity is proportional to vapour pressure (Figure 23). The absolute humidity of saturated air at 10, 20 and 30°C is approximately 8, 15 and 27 grams per kilogram respectively.

Saturation vapour pressure is the vapour pressure of water in equilibrium with a free water surface. Alternatively it can be defined as the pressure exerted by the maximum amount of water that can be contained in the air at a given temperature. The saturation vapour pressure increases rapidly as the air temperature rises.

Dew point is the temperature at which saturation occurs when air is cooled without change in water content. This is also a practical parameter, as it specifies temperature and 100 per cent relative humidity simultaneously, and hence saturation vapour pressure or saturation water content as well. Changes in air temperature above the dew point do not affect the water content, but cooling below the dew point removes moisture from the air by condensation on cooler surfaces. Moisture will condense on any surface cooler than the dew point of the air in contact with it.

Types of hygrometers

An instrument used to measure the amount of water in air is termed a hygrometer or psychrometer. Just as there are several ways of defining the amount of water in air, many methods have been devised for its measurement; no one hygrometer (or psychrometer), however, is suitable for all purposes over the full range of humidities and temperatures.

Wet and dry bulb hygrometers

This is the simplest and most widely used instrument for measuring humidity. It consists of two thermometers, one of which—the dry bulb—measures the air temperature. The wet bulb thermometer has a wet wick around the bulb. Evaporation of water from the wick into the atmosphere requires energy which comes from the remaining water and results in it being cooled. The drier the air, the greater is the rate of evaporation and hence the greater the depression of temperature. The temperature depression can be translated to per cent of relative humidity, water vapour pressure or dew point from tables prepared for this purpose. The values in the tables vary with atmospheric pressure, but the effect can be ignored for most practical purposes in the range 82 to 101 kilopascals (620 to 760 millimetres of mercury = altitudes to 1500 metres).

Certain precautions are necessary for accurate readings. The wick must be clean and free from dust and other contamination. Use of distilled water only is recommended. The bulbs must be ventilated with air moving at least 3 metres per second to ensure adequate evaporation and cooling of the wet bulb. The formula used to produce the tables assumes adiabatic evaporative cooling only, so the instrument should be protected from radiation sources, such as the sun, light bulbs or any surface much colder or warmer than the surrounding air. With careful operation and accurate thermometers reading to 0.1 deg. C, an accuracy of ± 1 per cent relative humidity is possible. The simplest is the sling (or whirling) psychrometer, while in the Assmann type, aspiration is by means of a small motorized fan. The latter is the standard instrument as it also has a built-in metal radiation shield and has finely calibrated and accurate thermometers. For general use a sling psychrometer is adequate and most practical. Wet- and dry-bulb instruments are now available with thermistors instead of mercury-in-glass thermometers; their advantages are small size, the possibility of automatic and distant operation, and remote control.

Hair hygrometers

These have as the sensing element several strands of hair, or a length of some other material that is capable of water sorption and desorption with a consequent change in length, and which are mechanically linked to a pointer on a scale. As the capillary sorption of water is slow, the response of these hygrometers is slow (10 to 30 minutes) and, as it also depends to some extent on the amount of water in the system, there are marked hysteresis effects. Therefore, the instruments should not be exposed to conditions of widely fluctuating temperatures or humidities. They must be calibrated for each temperature range. Over the range 30 to 80 per cent relative humidity they have an accuracy of 2 to 5 per cent relative humidity. They are useful for monitoring slow humidity variations at almost steady temperatures, as in cool stores.

Electric hygrometers

These instruments measure the psychrometric state of the atmosphere by recording variation in the resistance, capacitance or some other electrical parameter of a sensor with changing water sorption or desorption. Metal or carbon electrodes are attached to an insulating base impregnated with, or covered with a thin layer of, a dilute solution of an electrolyte, which equilibrates with the surrounding air by sorbing or desorbing water. The conduction types are temperature dependent and suffer from hysteresis and aging effects, but these instruments are small and have a fast response. The signals can be amplified and these types can be used readily for remote control. Capacitor sensors are more stable and reliable. They are unstable at high humidities, and must not be allowed to get wet.

Thin-film polymer hygrometers

Like some electric hygrometers, these measure changes in capacitance of a material that is sensitive to humidity. The sensor consists of a thin solid solution, which makes these instruments more robust and stable. Some can be washed with distilled water without losing calibration.

Dew point hygrometers

To measure the psychrometric state of the atmosphere with this type of hygrometer, air is cooled without change in water content until saturation is reached. The temperature (dew point) at which condensation is first seen on a cooled mirror surface is recorded: from this temperature the vapour pressure or relative humidity of air can be derived.

FURTHER READING

American Society of Heating. Refrigerating and Air-Conditioning Engineers. ASHRAE Handbook of fundamentals. Chapter 5. Psychrometrics. Chapter 12. Measurement and instruments. New York; 1972.

Gaffney, J.J. Humidity: basic principles and measurement. HortScience 13: 551–5; 1985.

Grierson, W.; Wardowski, W.F. Humidity in horticulture. HortScience 10: 356–60; 1975.

Henry, Z.A., ed. Instrumentation and measurement for environmental sciences. St. Joseph, MI: American Society of Agricultural Engineers; 1975.

Schurer, K. Comparisons of sensors for measurement of air humidity. Simatos, D.; Multon, J.L. eds. Properties of water in foods. Dordrecht: Martinus Nijhoff; 1985: 647–60.

Sharp, A.K. Humidity: measurement and control during the storage and transport of fruits and vegetables. CSIRO Food Res. Q 46: 79–85; 1986.

Szulmayer, W. Humidity and moisture measurement. CSIRO Food Preserv. Q. 29: 27–35; 1969.

Appendix VI

GAS ANALYSIS

Work with fresh produce frequently requires measurement of the concentrations of carbon dioxide, oxygen and ethylene. These gases can be measured by both chemical and physical methods. However, the latter methods are preferred because of their speed, accuracy and sensitivity. The method chosen depends on the available gas sample size, level of accuracy required, and whether continuous monitoring of gas concentration is necessary. Some techniques commonly used for gas analysis in postharvest work are briefly described.

Orsat gas analyser

The Orsat gas analyser is used for determining the concentrations of carbon dioxide and oxygen in large gas samples, such as those from controlled atmosphere storage rooms. The analyser consists of a calibrated burette connected by a glass manifold to two absorption tubes; the first tube is filled with potassium hydroxide to absorb carbon dioxide, the second with alkaline pyrogallol or ammoniacal cuprous chloride to absorb the oxygen. The concentration of carbon dioxide and oxygen is determined by the reduction in volume of the air sample as measured in the burette. An accuracy of 0.1 per cent is possible with the analyser.

Thermal conductivity gas chromatography

Thermal conductivity gas chromatography is particularly suited to the analysis of small samples, from 0.2 to 5 millilitres. Particular applications include the analyses of carbon dioxide and oxygen concentrations in small containers, for example, plastic bags, where the taking of samples of the size required for an Orsat gas analyser would drastically change the composition of the atmosphere in the bags.

Gas samples are fractionated generally on a silica gel column followed by a molecular sieve column into the component gases oxygen, nitrogen and carbon dioxide: the concentration of each component emerging from the columns is determined by a detector responsive to variations in the thermal conductivity of the separated gases passing over the detector.

Colorimetry

The colorimetric method of Claypool and Keefer is ideal for measuring carbon dioxide concentrations up to about 1.0 per cent. It is generally desirable to limit carbon dioxide to 0.5 per cent to prevent modified atmosphere effects on the produce. The method depends on maintaining a constant known flow of air over the sample and establishing equilibrium between the carbon dioxide in this air stream and a bicarbonate solution. The pH of this solution decreases as the carbon dioxide concentration increases. The change in pH can be measured colorimetrically using bromthymol blue. The original colorimetric procedure has been adapted to small laboratory spectrophotometers by Pratt and Mendoza.

Infrared analyser

This instrument measures carbon dioxide in flowing gas streams and is especially suitable for incorporation into automatic systems. Carbon dioxide absorbs infrared radiation at a specific wavelength, a property which is used to produce an electrical signal related to the carbon dioxide concentration in the test gas stream. Commercial instruments are not flow-sensitive.

Paramagnetic analyser

This instrument measures oxygen concentrations in flowing air streams. The paramagnetic, or magnetic susceptibility, analyser is limited to the analysis of oxygen and the oxides of nitrogen since these are the only paramagnetic gases, that is, gases attracted by a magnetic field. Paramagnetic analysers can determine differences in oxygen concentrations of 0.01 per cent.

Ethylene determination by gas chromatography

Gas chromatography using a flame ionization detector is a common, highly sensitive system for the measurement of ethylene and other hydrocarbons up to C5. The ethylene in the gas sample is separated from the other gases on an alumina column; the separated ethylene emerging from the column is mixed with hydrogen, burnt in air, and the ions given off in the flame provide an electrical signal proportional to the amount of ethylene present in the gas sample. Commercial instruments vary widely in sophistication, but a single column, single detector, instrument is adequate for routine ethylene analysis. The more sophisticated research instruments can measure 0.001 microlitres per litre ethylene in a 5 millilitre sample of air.

Other highly sensitive gas chromatograph detectors include the photo ionization detector. Carrier gas (usually nitrogen) only is required and the separated gases emerging from the column are fed directly to the detector.

FURTHER READING

Claypool, L.L.; Keefer, R.M. A colorimetric method for CO_2 determination in respiration studies. Proc. Am. Soc. Hortic. Sci. 40: 177–86: 1942.

Dilley. D.R.: Dewey, D.H.: Dedolph. R.R. Automated system for determining respiratory gas exchange of plant materials. J. Am. Soc. Hortic. Sci. 94: 138–41: 1969.

McNair. H.M.: Bonelli, E.J. Basic gas chromatography. Walnut Creek, CA: Varian: 1969.

Pratt, H.K.: Mendoza, D.B. Colorimetric determination of carbon dioxide for respiration studies. HortScience 14: 175–6: 1979.

Thompson, B. Fundamentals of gas analysis by gas chromatography. Walnut Creek. CA: Varian: 1977.

Watada, A.A.; Massie, D.R.A compact automatic system for measuring CO_2 and C_2H_4 evolution by harvested horticultural crops. HortScience 16: 39–41; 1981.

Young, R.E.: Biale, J.B. Carbon dioxide effects on fruit respiration. I. Measurement of oxygen uptake in continuous gas flow. Plant Physiol. 37: 409–15: 1962.

Index